Lecture Notes in Computer Science   **9064**

Commenced Publication in 1973
Founding and Former Series Editors:
Gerhard Goos, Juris Hartmanis, and Jan van Leeuwen

Editorial Board

More information about this series at http://www.springer.com/series/7410

Stefan Mangard · Axel Y. Poschmann (Eds.)

Constructive Side-Channel Analysis and Secure Design

6th International Workshop, COSADE 2015
Berlin, Germany, April 13–14, 2015
Revised Selected Papers

 Springer

Editors
Stefan Mangard
Graz University of Technology
Graz
Austria

Axel Y. Poschmann
NXP Semiconductors Germany GmbH
Hamburg
Germany

ISSN 0302-9743   ISSN 1611-3349 (electronic)
Lecture Notes in Computer Science
ISBN 978-3-319-21475-7   ISBN 978-3-319-21476-4 (eBook)
DOI 10.1007/978-3-319-21476-4

Library of Congress Control Number: 2015943438

LNCS Sublibrary: SL4 – Security and Cryptology

Springer International Publishing AG Switzerland is part of Springer Science+Business Media
(www.springer.com)

Preface

The 6th International Workshop on Constructive Side-Channel Analysis and Secure Design (COSADE) was held in Berlin, Germany, during April 13–14, 2015. This workshop each year brings together researchers and experts from academia, industry, and government who are working on cryptographic implementations and secure design.

COSADE 2015 received 48 submissions in the domain of side-channel analysis, fault attacks, and secure design, out of which 17 papers were selected. Each paper was reviewed by at least four independent reviewers. The Program Committee consisted of 33 members from 12 countries in America, Asia, and Europe who were carefully selected to represent a balanced view of both academia and industry. The members of the Program Committee were supported in their challenging task by 82 external reviewers. We would like to thank all committee members and reviewers for their hard work. The submission and reviewing process was done using the EasyChair system.

We were excited that Ross Anderson and Emmanuel Prouff accepted our invitations to give invited talks. Ross Anderson provided an excellent overview on "Why Cryptosystems Still Fail", while Emmanuel Prouff expounded on "Algorithmic Approaches to Defeat Side Channel Analysis". Beside the invited talks and the accepted papers, an update of the current state of the DPA Contest v4 was also presented at COSADE 2015. The paper "Side-Channel Security Analysis of Ultra-Low-Power FRAM-based MCUs" by Amir Moradi and Gesine Hinterwaelder received the best paper award.

We would like to thank the local organizers, in particular Claudia Petzsch and Matthias Petschik, as well as the general chair Jean-Pierre Seifert, for their support and for making this great event possible. On behalf of the COSADE community we would also like to thank the COSADE 2015 sponsors. Finally and most importantly, we would like to thank the authors for their excellent contributions.

May 2015

Stefan Mangard
Axel Y. Poschmann

Organization

Program Committee

Guido Bertoni	ST Microelectronics, Italy
Chien-Ning Chen	Nanyang Technological University, Singapore
Christophe Clavier	University of Limoges, France
Jean-Sébastien Coron	University of Luxembourg, Luxembourg
Elke De Mulder	Cryptography Research, USA
Hermann Drexler	Giesecke & Devrient, Germany
Thomas Eisenbarth	WPI, USA
Benoit Feix	Underwriters Laboratories, France
Wieland Fischer	Infineon Technologies, Germany
Benedikt Gierlichs	KU Leuven, Belgium
Christophe Giraud	Oberthur Technologies, France
Louis Goubin	Versailles University, France
Tim Güneysu	Ruhr-Universität Bochum, Germany
Johann Heyszl	Fraunhofer AISEC, Germany
Naofumi Homma	Tohoku University, Japan
Michael Hutter	TU Graz, Austria
Juliane Krämer	TU Berlin, Germany
Markus Kuhn	University of Cambridge, UK
Kerstin Lemke-Rust	Bonn-Rhein-Sieg University of Applied Sciences, Germany
Marcel Medwed	NXP Semiconductors, Austria
Amir Moradi	Ruhr-Universität Bochum, Germany
Debdeep Mukhopadhyay	Indian Institute of Technology Kharagpur, India
Elisabeth Oswald	University of Bristol, UK
Thomas Peyrin	Nanyang Technological University, Singapore
Emmanuel Prouff	ANSSI, France
Francesco Regazzoni	University of Lugano, Switzerland
Matthieu Rivain	CryptoExperts, France
Matt Robshaw	Impinj, USA
Akashi Satoh	University of Electro-Communications, Japan
Patrick Schaumont	Virginia Tech, USA
Alexander Schlösser	NXP Semiconductors, Germany
François-Xavier Standaert	UCL, Belgium

Additional Reviewers

Andouard, Philippe
Balasch, Josep
Basu Roy, Debapriya
Battistello, Alberto
Bauer, Sven
Belaid, Sonia
Bhasin, Shivam
Bilgin, Begül
Breier, Jakub
Chakraborty, Abhishek
Chen, Cong
Cioranesco, Jean-Michel
Daemen, Joan
Das, Poulami
De Santis, Fabrizio
Driessen, Benedikt
Durvaux, François
Endo, Sho
Farhady, Nahid
Feldhofer, Martin
Finiasz, Matthieu
Genelle, Laurie
Goodwill, Gilbert
Grosso, Vincent
Großschädl, Johann
Guo, Xiaofei
Hayashi, Yu-Ichi
He, Wei

Herrmann, Alexander
Heyse, Stefan
Hoffmann, Lars
Irazoqui, Gorka
Jap, Dirmanto
Jaulmes, Eliane
Kizhvatov, Ilya
Longo, Jake
Martinoli, Marco
Melzani, Filippo
Miele, Andrea
Montmasson, Julien
Okeya, Katsuyuki
Patranabis, Sikhar
Phuong Ha, Nguyen
Piscitelli, Robert
Poussier, Romain
Razafindralambo, Tiana
Rempel, Christof
Reparaz, Oscar
Roche, Thomas
Roussellet, Mylene
Saab, Sami
Samarin, Peter
Samotyja, Jacek
Sasdrich, Pascal
Schellenberg, Falk

Schläffer, Martin
Schneider, Tobias
Shaverdi, Aria
Shiozaki, Mitsuru
Stöttinger, Marc
Sugawara, Takeshi
Susella, Ruggero
Taha, Mostafa
Takahashi, Junko
Teglia, Yannick
Thillard, Adrian
Tibouchi, Mehdi
Tordella, Lucille
Tunstall, Michael
Ueno, Rei
van Niekerk, Eva
Varici, Kerem
Venelli, Alexandre
Verneuil, Vincent
Villegas, Karine
von Maurich, Ingo
Walle, Matthieu
Watanabe, Dai
Wild, Alexander
Wurcker, Antoine
Yuce, Bilgiday
Zankl, Andreas

Contents

Side-Channel Attacks

Improving Non-profiled Attacks on Exponentiations Based on Clustering and Extracting Leakage from Multi-channel High-Resolution EM Measurements

Robert Specht[1]([✉]), Johann Heyszl[1], Martin Kleinsteuber[2], and Georg Sigl[2]

[1] Fraunhofer Institute AISEC, Munich, Germany
{robert.specht,johann.heyszl}@aisec.fraunhofer.de
[2] Technische Universität München, Munich, Germany
{kleinsteuber,sigl}@tum.de

Abstract. The success probability of side-channel attacks depends on the used measurement techniques as well as the algorithmic processing to exploit available leakage. This is particularly critical in case of asymmetric cryptography, where attackers are only allowed single side-channel observations because secrets are either ephemeral or blinded by countermeasures. We focus on *non-profiled* attacks which require less attacker privileges and cannot be prevented easily. We *significantly improve* the algorithmic processing in *non-profiled* attacks *based on clustering* against exponentiation-based implementations compared to previous contributions. This improvement is mainly due to PCA and a strategy to select few mid-ranked components where exploitable, low-variance leakage is concentrated. As a result from a practical experiment using single-channel high-resolution magnetic field measurements, we report a significant improvement in the number of successful attacks. Further, we present the first practical results from using three such channels simultaneously. The combination of three channels leads to further improved results over the best individual channel when applying a *profiled* template attack. The *clustering-based* algorithmic approach for the *non-profiled* attack, however, does not show improvements from the combination.

1 Introduction

The side-channel information leakage about secret-dependent internal values is usually limited. Attackers who target implementations of symmetric ciphers may repeat measurements many times to collect sufficient leakage information while the secret remains unchanged. In case of asymmetric algorithms, however, the secret is either ephemeral or blinded through countermeasures and attackers are only allowed one side-channel observation. Hence, it is crucial to *record and exploit* as much leakage as possible. *Profiled* attacks, e.g. template attacks, are powerful in exploiting leakage efficiently, however, can be prevented by blinding or by preventing attackers from gaining full access for profiling. *Non-profiled*

© Springer International Publishing Switzerland 2015
S. Mangard and A.Y. Poschmann (Eds.): COSADE 2015, LNCS 9064, pp. 3–19, 2015.
DOI: 10.1007/978-3-319-21476-4_1

attacks cannot be prevented in this way, because they do not require profiles and hence, are a much bigger threat to devices. Heyszl et al. [8] proposed to use *well-established* clustering algorithms for *non-profiled* attacks. They use k-means clustering after a simple sum-of-squares pre-processing of the measurement data in their practical experiments.

We follow their proposal and *significantly improve* the algorithmic approach. Principal Component Analysis (PCA) has been used for pre-processing and data reduction in other side-channel attacks [2,3,5,15,22]. Also, strategies to select only certain principal components have previously been mentioned [3]. We apply (PCA) to *clustering-based, non-profiled* attacks on exponentiation algorithms and performed practical experiments on an FPGA-based implementation of Elliptic Curve Cryptography (ECC) by using high-resolution electromagnetic measurements as side-channel. We find that PCA concentrates exploitable leakage with comparably low variance into few components which are not the highest-ranked ones. Hence, as an important step after transformation, we discard high-ranked as well as many low-ranked components during a parametrized selection. In our *non-profiled* setting, this requires some testing for the right selection parameters, hence, brute-force by the attacker. However, significantly improved attack results clearly justify this. For cluster classification, we use the expectation maximization algorithm instead of the k-means algorithm [8]. The resulting attack is successful with single-channel measurements in significantly more cases than if using the algorithmic approach by Heyszl et al. [8]. Most of the achieved algorithmic improvement can be attributed on using PCA and the component selection as pre-processing technique before clustering. Like expected, a *profiled* template attack still outperforms the improved *non-profiled* attack.

Another way to improve attacks in single-execution settings is to use *multiple simultaneous channels* and combine their leakage. Previous contributions have tested the combination of (low-resolution) magnetic field measurements and current consumption measurements [1,22] using template attacks. High-resolution magnetic field measurements should generally provide better signal qualities [10] and allow to capture multiple independent channels because the signals highly depend on measurement locations [9]. We present the *first practical results from using three high-resolution magnetic field probes simultaneously* and combine them in the clustering-based non-profiled attack. However, we find that the combination of three channels does *not improve* the results using the *non-profiled PCA- and clustering-based* attack compared to the best individual channel. We conclude that in the *non-profiled* setting, our approach seems unsuitable for combining multi-channel data. The *profiled* template attack, however, leads to a *significant improvement* through the combination of channels. In profiled settings, attackers are able to find the best measurement positions for single channels. Hence, the additional cost for multi-channel equipment is only reasonable in *profiled* settings *and* if the available *leakage* is still *insufficient*.

We first explain the background and related work of (non-profiled) attacks against exponentiations in Sect. 2.1. In Sect. 2.2, we cover the background and related work of magnetic field side-channels and multi-channel measurements.

In our first main Sect. 3, we describe our algorithmic approach to improve clustering-based attacks on exponentiations and to handle multi-channel data. We back these considerations by practical experiments in Sect. 4 and discuss the results. We summarize our contribution and findings in Sect. 5.

2   Preliminaries

2.1   Non-profiled Attacks Against Exponentiations

The main computation in public key cryptosystems is modular integer exponentiation with secret exponents (e.g. RSA, DSA) or elliptic curve scalar multiplication (e.g. ECDSA) with secret scalars. In this contribution, we use the generalized terms 'exponentiation algorithms' and 'secret exponents'. The secret exponent is usually either ephemeral by design (e.g. ECDSA) or blinded through countermeasures (e.g. exponent blinding in RSA, or in ECDSA to prevent profiling). Therefore, it is different for every execution and side-channel attackers may only exploit *single executions*. The first *single-execution* attack on exponentiations was presented by Kocher [13] who exploits data-dependent execution times of algorithms. To avoid this, improved algorithms like the square-and-multiply-always, double-and-add-always or the Montgomery ladder algorithm have constant operation sequences (e.g. side-channel atomic routines) to avoid such *simple side-channel attacks*. In all those algorithms, exponents are scanned bit- or digit-wise (depending on whether it is a binary, m-ary, or sliding window exponentiation) and the computation is performed in a loop iterating a constant sequence of operations. (We will continue to refer to the binary case in this contribution.) Nonetheless, some side-channel leakage about the processed exponent remains in many cases which can be referred to as single-execution leakage. Examples include data-dependent leakage from using pre-computed multiples in digit-wise multiplications [25], address-bit leakage [12], location-dependent leakage from accessing different storage locations [9], or operation-dependent leakage, e.g., when square and multiply operations can be distinguished [4].

Attacks against an exponentiation are carried out by partitioning side-channel measurements into *trace-segments* with each segment corresponding to an independently processed bit of the secret exponent. The segmentation borders are either known a priori, or can often be derived from visual inspection or comparison of shifted trace parts. The trace for measuring n exponent bits consists of n trace-segments $\boldsymbol{t_d} = (t_{1+(d-1)\cdot l}, ..., t_{d\cdot l})$ with $d \in [1, n]$, each of which is of length l (time-samples) which is referred to as its dimensionality (of features). For analyzing and attacking the measurement data, a $n \times l$ matrix $\boldsymbol{M}$ is constructed by placing each segment in one row. The contained leakage is exploited to find a structure, or partitioning of the rows due to secret exponent values. Template attacks use a profiling step to create templates of the segments for different values. Profiling can be prevented in many cases by blinding countermeasures or not allowing attackers full access to devices for profiling. We concentrate on *non-profiled* attacks because they are more powerful and threatening.

There have been several published attacks on exponentiations which do *not* require profiling. Walter [25] was the first to describe an attack by using a *custom* algorithm (resembling a clustering algorithm) to partition the segments into buckets. Messerges et al. [16], Clavier et al. [6], and Witteman et al. [26] use cross-correlation in non-profiled single-execution attacks on exponentiations. We pursue the approach by Heyszl et al. [8] who promote the use of established clustering algorithms (such as e.g. *k-means*) for *non-profiled* attacks due to the generality of their approach and support for the combination of multiple channels. A correct classification of trace-segments equals the recovery of the secret exponent. (Later, Perin et al. [18] described a similar but heavily customized two-stage approach which seems tailored to their case and unreasonable for generalization.) *We extend and significantly improve previous work by using Principal Component Analysis (PCA) and expectation maximization clustering (instead of k-means and simple pre-processing).*

2.2 Multi-probe Measurements of Magnetic Fields

Using multiple side-channels concurrently, and combining them in an attack is an important way of increasing the exploitable leakage in single-execution attacks. Agrawal et al. [1] first, and later Standaert et Archambeau [22], describe the combination of *current consumption* with *magnetic field* measurements in *profiled* attacks through *concatenation* of traces. Standaert and Archambeau [22] report better results from magnetic field than current measurements and report an improvement from the combination of both channels. Souissi et al. [21] first presented results from combining *two* simultaneous measurements of the *magnetic field*. They measure the field close to two different supply capacitors of an FPGA. In this way they measure the supply of two different parts of the FPGA.

We find that in many cases, side-channel measurements of the magnetic field are closely related to the consumption of an *entire* device because comparably large coil diameters (>500 um) are used at large distances to the integrated circuits (>300 um) [1,7,17,20,22]. Such measurements often capture the magnetic field of supply wires (bonding wires) which is directly proportional to the current consumption of the *entire* integrated circuit (including noise sources from within the device). In our opinion, it is *unreasonable* to simultaneously record more *than one magnetic field* channels in such cases due to this global character. Lately, high-resolution magnetic field measurements at close distances to an integrated circuit die have been investigated extensively by Heyszl et al. [9,10]. Such high-resolution measurements require magnetic field probes with diameters of $\approx 150\,\mu$m at close distances to an integrated circuit die (<100 μm). In our opinion, the capturing of *multiple simultaneous magnetic field* side-channels *only* makes sense in case of such *high-resolution* measurements which can be restricted to *parts* of integrated circuits because they will convey sufficiently different information (e.g. localized leakage [9]). Heyszl et al. [8] mention the combination of multiple high-resolution channels for *non-profiled single-execution* attacks, however, did not perform actual simultaneous measurements. *We extent their work and present first results from an extensive practical study using three* high-resolution *micro-coil magnetic field channels.*

3 Improving Clustering-Based Attacks

In this section, we describe our algorithmic approach to clustering-based *non-profiled* attacks on exponentiations which improves previous work [8]. We explain how we use Principal Component Analysis (PCA) as a pre-processing step for dimensionality reduction and feature selection in Sect. 3.1. We continue and describe how expectation maximization clustering can be used to attack single- and multi-channel measurements in Sect. 3.2. Finally, we describe how classification errors can be handled and derive the brute-force complexity as a measure to assess attack outcomes in Sect. 3.3.

3.1 PCA for Dimensionality Reduction and Feature Selection

Side-channel measurements usually lead to big amounts of data, especially when high sampling rates for magnetic field measurements are required. This increases required computational power and memory consumption during subsequent data analysis. Only a small part of the data will contain exploitable leakage information. Hence, *feature selection* to discard other parts is desirable.

Simple trace compression [14] is commonly used and usually justified by electrical properties. This includes extracting the peak values or computing the sum-of-squares (such as Heyszl et al. [8]) during the time-period of one clock cycle. Another popular method is the selection of so-called points-of-interest. This subset is usually identified through profiling with known secrets.

We concentrate on powerful *non-profiled, unsupervised* methods, specifically, on PCA. PCA has been applied to side-channel analysis for data reduction in several contributions [2,3,5,15,22] for different attacks of which Archambeau et al. [2] were the first to describe the use of PCA in the context of template attacks. Standaert and Archambeau [22] later compare PCA and Linear Discriminant Analysis (LDA) in the context of template attacks and confirm that LDA leads to superior results. We disregard LDA because training data from profiling is used to achieve a representation which maximizes cluster separation.

PCA is based on Singular Value Decomposition (SVD) and transforms the data into another coordinate system subspace with linearly uncorrelated coordinates by using the variance as score function, hence, maximizing the retained variance of the data. As described in Sect. 2.1, recorded side-channel measurements are cut into trace-segments corresponding to exponent bits. This leads to the real matrix M of measurement data, with the shape $n \times l$ for every probe (see Sect. 2.1). The SVD of $M = U * \Sigma * V^*$ and the transformation into the orthogonal subspace of M equals $U * \Sigma$. This matrix $U * \Sigma$ consists of column vectors $(PC_1, ..., PC_r)$ with r being the number of row-vectors and PC_j being a column-vector of shape $n \times 1$, which is called a principal component. The maximum number of components equals the number of trace-segments n of the original data, $\max |PC| = \min(n, l)$, because the segment-length l is usually much larger than n. After applying PCA, the components are ordered by their variance which can be found in the diagonal matrix Σ. In our experiments, we normalize the variances of the principal components to one, i.e. we directly use $M_{\mathrm{PCA}} = U$ instead of $U * \Sigma$. Before applying PCA, we removed the mean of every trace-segment as a standard measure.

Ideally, a transformation into a reduced subspace should maintain the 'useful' information while neglecting 'not useful' information, which is difficult without supervision. PCA combines correlating input dimensions into single principal components. Archambeau et al. [2] propose to only retain the first-ranked components assuming that the leakage is contained there, while discarding the remaining low-variance ones, assuming only noise is contained. Batina et al. [3] found in their practical experiments, that results of correlation-based Differential Power Analysis (DPA) improved when removing first-ranked components. There are several reasons for high variances of the trace segments, e.g. data-dependent signal influences and noise, which are irrelevant to the desired classification. We suspect that relevant and irrelevant signal parts will aggregate within separate components. Also, from our experience, the 'interesting' leakage signal parts are rather low-variance in the case of single-execution attacks.

Hence, we *propose a selection strategy* which discards several highest-ranked as well as many low-ranked components because they either contain noise or information which we are not interested in. We either select *single* principal components or a number of *consecutive* components (random choices of multiple components will lead most likely to an untestable amount of possibilities). Reduced trace-segments $M_{\mathrm{PCA},k:k+i} = (PC_k, ..., PC_{k+i})$ are derived with k the first selected component and $i \geq 1$ the number of consecutive components retained. We trialled values of $k \in [1, 20]$ and $i \in \{1, 2, 4, 6, 9\}$ in our practical experiments and found that using only one single component $i = 1$ leads to the best results in our attack on average, and that the $k \leq 3$ first-ranked components should be discarded. This selection strategy reflects the approach of an attacker who is unable to perform profiling. An optimal selection of components can certainly *not* be determined *a priori* because it is highly device- and application-specific (general issue in machine learning [27]). Hence, without a priori-knowledge, an attacker has to trial different values for k and i. This, however, only requires an additional *brute-force complexity* of a few bits and improved attack outcomes clearly justify this.

3.2 Expectation-Maximization Clustering of Multi-channel Data

Clustering algorithms can generally be split into supervised, semi-supervised and unsupervised algorithms. Our focus on *non-profiled* attacks restricts the choice to *unsupervised* algorithms. Heyszl et al. [8] first describe how *unsupervised* clustering algorithms can be used in a *non-profiled* attack to partition n trace-segments into classes according to their secret exponent values. An unsupervised cluster classification is equivalent to estimating the free parameters of the classes' assumed distribution model. The choice of the algorithm and free parameters depends on the assumed probability distribution model, hence shape of the clusters. While Heyszl et al. use k-means clustering, we improve this by using the expectation maximization algorithm while keeping the Gaussian distribution assumption which both algorithms are based on.

Expectation-maximization clustering provides more free parameters which leads to a generally improved approximation of the cluster distributions, which usually leads to better classification results. The algorithm is based on repeated expectation and maximization steps. During these iterations the maximum likelihood means and covariances for the Gaussian distribution are derived. The result is a classification and a class-membership probability which indicates the reliability of correct classification for each segment (resp. secret exponent bit). The number of free parameters in the clustering algorithm can be chosen. We assume that the cluster shapes are mainly defined by Gaussian distributed noise. Additionally, we assume the noise being independent of the processed bit value. Hence, we chose to estimate two means and one joint full covariance matrix.

Multiple simultaneous measurements channels are combined by concatenating the trace-segments from different channels which correspond to the same exponent bits [1,8]. PCA is applied to all side-channel measurement channels separately before concatenation. For example, segments $M^1_{\mathrm{PCA},k:k+i}$ from measurement channel 1 are combined with segments $M^2_{\mathrm{PCA},k:k+i}$ from measurement 2 leading to combined segments $M^{\mathrm{combined}}_{\mathrm{PCA},k:k+i} = (M^1_{\mathrm{PCA},k:k+i}, M^2_{\mathrm{PCA},k:k+i})$. An attacker would rather use the *same values for k and i in all channels* because it significantly increases the attack complexity to test different k-s and i-s for every channel *without profiling* (e.g. repeat clustering process $(20 * 5)^3$ times).

3.3 Classification Errors and Required Brute-Force Complexity

If the recovered exponent is incorrect, faulty bits need to be identified, which is usually hard. As described by Heyszl et al. [8], an attacker can use the bits' probabilities of correctness to judge which need to be trialled for correctness and follow a simple strategy to enumerate possible keys. This strategy leads to an estimated remaining *brute-force complexity* which we use to assess practical attack outcomes. Better, even optimal, key enumeration strategies [23,24] will result in a lower amount of required brute-force *if* the attacker applies them. However, the typically large key sizes in asymmetric cryptography make the application of such algorithms challenging for attackers as well as evaluators. The said *brute-force complexity* which is used instead can be seen as an upper bound for the rank of the correct key as derived from an optimal enumeration.

We chose to use the silhouette index score [19] for the bits' error probability. It is based on the cumulative distance of each trace-segment to other trace-segments of each cluster. The silhouette index is calculated for every m_{PCA}, which corresponds to one row of $M_{\mathrm{PCA},k:k+i}$, with C_1 being the set of trace segments t_d of the same cluster like m_{PCA} (determined by the expectation maximization algorithm) and C_2 being the set of trace segments belonging to other clusters. With the distance function $dist(a, b)$ (we use Euclidean distance due to the Gaussian noise assumption) the silhouette index s is computed as:

$$s(\boldsymbol{m}_{\text{PCA}}, \mathcal{C}_1, \mathcal{C}_2) = \frac{f(\mathcal{C}_1, \boldsymbol{m}_{\text{PCA}}) - f(\mathcal{C}_2, \boldsymbol{m}_{\text{PCA}})}{max(f(\mathcal{C}_1, \boldsymbol{m}_{\text{PCA}}), f(\mathcal{C}_2, \boldsymbol{m}_{\text{PCA}}))} \qquad (1)$$

$$f(\mathcal{C}, \boldsymbol{m}_{\text{PCA}}) = \frac{1}{|\mathcal{C}|} \sum_{x \in \mathcal{C}} dist(x, \boldsymbol{m}_{\text{PCA}}) \qquad (2)$$

After calculating the score for all n segments, the ones with the lowest s are brute-forced in repetitions while including an increasing number of bits [8]. Let q be the last bit which is trialled until the correct exponent is found, then $2^{(q+1+1)}$ different exponents have to be tested at maximum which can be referred to as remaining *brute-force complexity* after the attack [8]. One additional bit is included for both possibilities to assign labels to the two classes. It equals $2^{(n+1+1)}$ at maximum and 2^1 at minimum.

4   Practical Evaluation

We present the first practical results from the *simultaneous* use of *three high-resolution* magnetic field probes. We chose a fixed geometric arrangement of the measurement probes close to the surface of an FPGA die and performed 400 measurements at different positions to gain conclusive insights from a high number of tests. We succeed in *demonstrating the algorithmic improvement* from our approach and derive conclusions about the *benefit from using multiple channels simultaneously*.

4.1   Design-Under-Test and Multi-probe Setup

As a device under test, we use a Xilinx Spartan 3A FPGA chip (see Fig. 1a) which is configured with an Elliptic Curve Cryptography (ECC) design and performs an 163 bit elliptic curve scalar multiplication using a Montgomery ladder. This algorithm is a classical candidate for attacks against exponentiation algorithms

   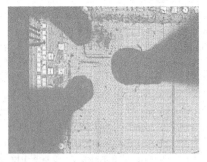

(a) View from side   (b) Top-view

Fig. 1. Geometric arrangement of measurement-probes on FPGA die surface

since it processes the secret exponent bit-wise in n constant time segments. As a single-execution side-channel leakage about the consecutively processed exponent bits we exploit location-based leakage which is revealed by high-resolution measurements of the electromagnetic field [9].

After decapsulating the FPGA die (see Fig. 1a), we use an area of $1700\,\mu m \times 1700\,\mu m$ on the surface of the die between bonding wires to place probes. We arrange three probes in a fixed formation, and place them on 400 (20×20) different positions within this area to able to evaluate 400 data sets by our analysis. Figure 1 depicts the geometric arrangement of the probes from the side and from the top. The distance of the probes to the die surface is approximately $100\,\mu m$. We used three near-H-field (magnetic) probes with coil diameters of $250\,\mu m$, $150\,\mu m$ and $100\,\mu m$ which we had available in our laboratory. The bandwidth of the probes is $6\,GHz$ with a built-in $30\,dB$ amplifier. The signal is sampled synchronously to the device's clock at $2.5\,GS/s$. Contrary to other contributions [8,18] no simple compression or pre-proccessing (e.g. averaging, maximum extraction or sum-of-squares during clock cycles) is applied before (PCA) and clustering. Such simple trace pre-processing techniques have been shown to have negative effects on results [10].

4.2   Quality of Principal Components

Our algorithmic approach includes the selection of principal components after PCA as a first step before clustering. The selection can be described by two parameters, k the first selected component, and, $i \geq 1$ the number of consecutively selected components after the k-th one as described in Sect. 3.1. In this section we investigate the quality of different parameter choices. We executed the clustering-based attack on *every* single measurement from all 3 probes and 400 positions with choices of $k \in [1, 20]$ and $i \in \{1, 2, 4, 6, 9\}$ and assess the quality using the remaining brute-force complexity explained in Sect. 3.3.

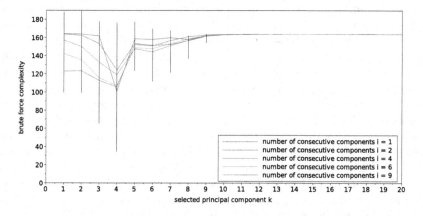

Fig. 2. Mean brute-force complexity for different selected principal components (k and i) over all measurement positions *including standard deviation as bars*

We show the means over $3 * 400$ results for the resulting brute-force complexities for each combination of parameters k and i in Fig. 2. Hence we are able to equally compare the results of different probes and show some fundamental properties of our measurements. These high *mean* brute-force complexities of >100 bits are certainly not within the range of realistic computing capabilities. They result from including many low-scoring results. The standard deviations are shown as vertical bars and indicate that there are multiple results with significantly lower brute-force complexities (the diagram does not include +1 bits for assigning labels to classes). As an important observation, low-ranked components ($k < 10$) seem preferable overall and *first-ranked principal components do not contain exploitable leakage* (see curve with $i = 1$ or $i = 2$ in Fig. 2). This confirms our assumptions from Sect. 3.1 as well as similar observations from Batina et al. [3]. Thus, we *discard first-ranked as well as low-ranked principal components* before further analysis and achieve significantly improved brute-force complexities.

In Fig. 2 it can also be noted that the component number $k = 4$ seems to contain the most leakage on average, reaching the lowest mean brute-force complexities. It seems that PCA *concentrates most of the exploitable leakage information into a single principal component*. This means that a choice of $i = 1$ for the number of selected consecutive principal components led to the best results in our circumstances. We used this choice in the practical evaluation in the next Sect. 4.3. As another observation, curves with $i > 2$ lead to low complexities as soon as component 4 is included in the consecutively selected components. For illustrative purposes, we show the resulting principal components after PCA transformation of an examples trace in the Appendix A.1.

4.3   Analyzing Separate Channels

For every probe, we have 400 measurements from different positions. We analyze the data from the three available channels separately: Firstly we perform preprocessing by applying PCA, secondly we perform clustering using the expectation maximization algorithm and thirdly we compute the remaining brute-force complexity. For every probe separately, and for every selection of principal components (for every $k \in [1, 20]$ while $i = 1$), we summarize the results from 400 tests in Fig. 3a, 3b, and 3c. Figure 3a shows results for the 250 μm coil probe, Fig. 3b for the 150 μm coil probe and Fig. 3c for the 100 μm coil probe. The figures show, *how many of the* 400 *measurements of each probe*, and for every selection of k, lead to which brute-force complexities. The occurrence rate is visually indicated by the size of the respective dots. Bigger dots mean that the corresponding brute-force complexity has occurred more often. For example, in Fig. 3a, almost all of the 400 measurements lead to a maximum brute-force complexity of 163 for $k < 5$ and $k > 10$. For $k = 5$, however, many measurements lead to lower resulting brute-force complexities, some even of the minimum. The red dashed line highlights the 32 bit complexity level up to which all outcomes are *easily manageable* for attackers through computation.

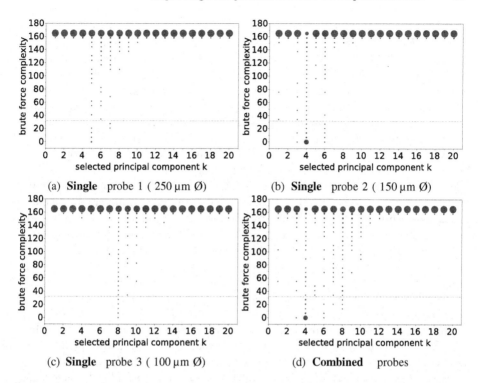

Fig. 3. Brute force complexity occurrences over different principal components (Color Figure Online)

As an important finding, it can be observed, that the *probe* with the 150 μm coil diameter depicted in Fig. 3b leads to the *best results* by far. For the principal component $k = 4$, *an astonishing percentage of* 56 % out of the 400 measurements led to a remaining brute-force complexity ≤ 32 bit (summing up all outcomes equal or lower the red dashed line). This high number was unexpected and means that *with the improved algorithmic, more than half of all measurement positions exhibited sufficient leakage* for a complete break. The 100 μm probe depicted in Fig. 3a leads to only 3 % ≤ 32 bit for $k = 5$. Also the 250 μm probe depicted in Fig. 3c only leads to 3 % ≤ 32 bit for $k = 8$. Hence, the 150 μm coil probe seems to work best under our circumstances. Since finding suitable measurement positions is rather easy (using the best probe), attackers should test different measurement positions instead of employing extensive computational brute-force, testing is comparably easy in case of single-execution attacks because only single measurements need to be analyzed at every position.

Without knowing $k = 4$ and $i = 1$ a priori, attackers could make minimal heuristic assumptions like $k \in [3, 10]$ and $i \in [1, 4, 9]$ which could fit similar circumstances. This would result in an additional brute-force complexity of $+4$ bits which is not included in Fig. 3 and justified by significantly improved results.

To demonstrate the improvement of our proposal, we performed the *original attack of Heyszl et al.* [8] on the *same measurements*. A remaining brute-force complexity ≤32 bit is reached in *none* (0 %) of the 400 measurement cases using the 150 μm coil probe. Compared to 56 % from the improved attack, this means that we achieve *astonishingly improved results* from applying PCA and expectation maximization clustering. (Only the 250 μm coil probe led to marginally better results using the previous method, i.e., 8 % instead of 3 % of the cases ≤32 bit, however, this does not invalidate the previous statement in our opinion.)

We compared the performance of the *k*-means versus the expectation maximization clustering algorithm in the context of single channels. Since we only select single components ($i = 1$) after PCA, channels only consist of single dimensions and there is not much benefit from more free parameters in the clustering algorithm. This is confirmed by the fact that expectation maximization and *k*-means clustering *lead to almost equal results*. This means that our reported *improvement is mainly due to the PCA transformation and the selection of components*. In the multi-channel case, however, more dimensions aggregate from separate channels making expectation maximization more eligible.

As *benchmark* for *high-resolution magnetic field measurements*, we tested the improved *non-profiled* attack on a *current consumption measurement*. We use a 1 Ohm measurement resistor and a differential probe at unchanged sampling rate. To cancel one-time effects such as disturbances or noise, we repeated this 12 times and averaged the results. The outcome is a *significantly high brute-force complexity of* 152 *bits*. Hence, it is completely impossible to exploit leakage from such current measurements. This underlines that high-resolution magnetic field measurements are clearly *superior* in leakage signal quality in our circumstances.

4.4   Combining Multiple Channels

After the individual analysis of the three measurement channels, we combined the channels for analysis as described in Sect. 3.2. The motivation for attackers to combine channels is to increase the exploitable leakage to improve attack outcomes, e.g. instead of trying to find better measurement positions.

Figure 3d shows the brute-force complexity results for the combined measurements in the same way as described in the previous Sect. 4.3. A *visual comparison* of the combined results in Fig. 3d to the individual results in Fig. 3a, b, and c gives the impression, that the overall result is comparable to Fig. 3b. However, expressed quantitatively in the same way as before, the combined channels lead to a remaining brute-force complexity of ≤32 bit in only 52 % of the cases for $k = 4$. Hence, as an important result, *instead of an improvement*, we observe a *slight degradation* compared to the best individual case which led to 56 % of cases ≤32 bit. This means that the described clustering-based *non-profiled* attack is *unable to benefit* from a combination of channels (in our circumstances).

We suspect that this is due to the fact that our selection strategy selects equal values k and i to pre-process all three channels in the same way using PCA in case of combined attacks. This should be a significant disadvantage in our case where different k are best for different channels (see results in Sect. 4.3).

Unfortunately, it would significantly increase the complexity to test different k-s for every channel (e.g. repeat clustering 20^3 times). Increasing the number of selected components i to prevent this would include more noise, in our circumstances, which in turn would degrade classification results significantly (see how curves with $i > 1$ result in higher mean-values in Fig. 2).

We compared the improved *non-profiled* attack against a *profiled* template attack. This requires one additional trace for profiling at each position and for every probe. Templates consist of two means and a single full covariance matrix. To derive the remaining brute-force complexity as described in Sect. 3.3, we use bit-wise template matching results. For a fair comparison, we also apply PCA including the selection strategy for k and i. A *higher number of* 61 % of positions (compared to the 56 % from the non-profiled attack) lead to remaining brute-force complexities ≤ 32 bit for the $150\,\mu$m coil probe, with $i = 1$ and $k = 4$. The *profiled* template attack outperforms the *non-profiled* attack. As the most important observation, we find that the *combination of channels leads to an improved* 66 % of the cases with a remaining brute-force complexity of ≤ 32 bit, with $i = 9$ and $k = 3$. This clearly demonstrates *the gain of combining channels in the profiled setting.*

In a *profiled setting*, attackers are able to test and find the 'best' measurement positions. This means that, in our circumstances, the use of multiple channels is only reasonable if the leakage of such best single channels is insufficient which diminishes the good results to a certain extent.

5 Conclusion

We significantly improved the algorithmic approach for *non-profiled* attacks against exponentiation by applying (PCA) and disregarding high- as well as low-ranked ones following a simple strategy. This selection strategy requires some trying-out (additional brute-force), but this is highly rewarded by improved attack outcomes in terms of low brute-force complexities. With this approach, the unsupervised attack using a single-channel high-resolution magnetic field measurement is remarkably threatening and leads to manageable brute-force levels in over half of the tested measurement positions. This emphasizes the need to prevent *all possible cause for exploitable single-execution leakage.* Regarding our results from three simultaneous channels, we find that the combination of channels only significantly improves the attack results, if a *profiled* attack is used. In case of the *clustering-based, non profiled* attack, the results from the combination are only comparable to the best individual one. In profiled settings attackers are also able to look for the 'best' measurement positions. Hence, multi-channel attacks are only reasonable if the exploitable leakage is insufficient at such best positions.

Acknowledgements. This work was partly funded by the German Federal Ministry of Education and Research in the project SIBASE through grant number 01IS13020.

A   Appendix

A.1   Illustration of Principal Components After Transformation

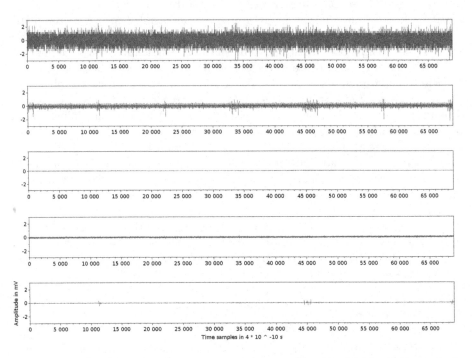

Fig. 4. Example of an original trace-segment (topmost) and its high-ranked principal components below. The 4-th (bottom) component contains signal leakage

Figure 4 depicts principal components after Principal Component Analysis (PCA) transformation for illustrative purposes. We used an example measurement where the side-channel leakage is sufficient for the attack to succeed without false classifications when selecting the $k = 4$-th component for expectation maximization clustering. The topmost diagram depicts one trace-segment in its original form. Below this, the four highest-ranked principal components of this segment are depicted. From the previous analysis we know that the exploitable leakage seems concentrated in component $k = 4$ which is depicted in the bottom diagram. The time-samples with higher values represent the times of exploitable leakage information in this component. The sparse occurrence fits to the description of data-dependent register accesses as source of this leakage [9]. A comparison to the other components in Fig. 4 clearly shows that the *leakage is small compared to the remaining signal parts*.

A.2 Countermeasures

As previously described by Heyszl et al. [8], countermeasures such as exponent blinding do not protect against *non-profiled* attacks. Many countermeasures address individual single-execution leakage sources of implementation (e.g. address-bit, or localized leakage).

As a conclusion from this contribution, we must emphasize the necessity to reduce all possible single-execution leakage sources as much as possible.

Homma et al. [11] present a general countermeasure against high-resolution magnetic field measurements. They describe an on-chip sensor which detects magnetic field probes in close distance to die surfaces. However, in our opinion this will not help since measurement probes are typically placed close to an integrated circuit before power-up. Hence, necessary calibration routines of the sensor will likely not be able to distinguish the static probes from other environmental influences.

References

1. Agrawal, D., Rao, J.R., Rohatgi, P.: Multi-channel attacks. In: Walter, C.D., Koç, Ç.K., Paar, C. (eds.) CHES 2003. LNCS, vol. 2779, pp. 2–16. Springer, Heidelberg (2003)
2. Archambeau, C., Peeters, E., Standaert, F.-X., Quisquater, J.-J.: Template attacks in principal subspaces. In: Goubin, L., Matsui, M. (eds.) CHES 2006. LNCS, vol. 4249, pp. 1–14. Springer, Heidelberg (2006)
3. Batina, L., Hogenboom, J., van Woudenberg, J.G.J.: Getting more from PCA: first results of using principal component analysis for extensive power analysis. In: Dunkelman, O. (ed.) CT-RSA 2012. LNCS, vol. 7178, pp. 383–397. Springer, Heidelberg (2012)
4. Bauer, S.: Attacking exponent blinding in RSA without CRT. In: Schindler, W., Huss, S.A. (eds.) COSADE 2012. LNCS, vol. 7275, pp. 82–88. Springer, Heidelberg (2012)
5. Bohy, L., Neve, M., Samyde, D., Quisquater, J.J.: Principal and independent component analysis for crypto-systems with hardware unmasked units. In: Proceedings of e-Smart (2003)
6. Clavier, C., Feix, B., Gagnerot, G., Roussellet, M., Verneuil, V.: Horizontal correlation analysis on exponentiation. In: Soriano, M., Qing, S., López, J. (eds.) ICICS 2010. LNCS, vol. 6476, pp. 46–61. Springer, Heidelberg (2010)
7. De Mulder, E., Örs, S.B., Preneel, B., Verbauwhede, I.: Differential power and electromagnetic attacks on a FPGA implementation of elliptic curve cryptosystems. Comput. Electr. Eng. **33**, 367–382 (2007)
8. Heyszl, J., Ibing, A., Mangard, S., De Santis, F., Sigl, G.: Clustering algorithms for non-profiled single-execution attacks on exponentiations. In: Francillon, A., Rohatgi, P. (eds.) CARDIS 2013. LNCS, vol. 8419, pp. 79–93. Springer, Heidelberg (2014)
9. Heyszl, J., Mangard, S., Heinz, B., Stumpf, F., Sigl, G.: Localized electromagnetic analysis of cryptographic implementations. In: Dunkelman, O. (ed.) CT-RSA 2012. LNCS, vol. 7178, pp. 231–244. Springer, Heidelberg (2012)

10. Heyszl, J., Merli, D., Heinz, B., De Santis, F., Sigl, G.: Strengths and limitations of high-resolution electromagnetic field measurements for side-channel analysis. In: Mangard, S. (ed.) CARDIS 2012. LNCS, vol. 7771, pp. 248–262. Springer, Heidelberg (2013)

11. Homma, N., Hayashi, Y., Miura, N., Fujimoto, D., Tanaka, D., Nagata, M., Aoki, T.: EM attack is non-invasive? - design methodology and validity verification of EM attack sensor. In: Batina, L., Robshaw, M. (eds.) CHES 2014. LNCS, vol. 8731, pp. 1–16. Springer, Heidelberg (2014)

12. Itoh, K., Izu, T., Takenaka, M.: Address-bit differential power analysis of crypto-graphic schemes OK-ECDH and OK-ECDSA. In: Kaliski, B.S., Koç, Ç.K., Paar, C. (eds.) CHES 2002. LNCS, vol. 2523, pp. 399–412. Springer, Heidelberg (2003)

13. Kocher, P.C.: Timing attacks on implementations of Diffie-Hellman, RSA, DSS, and other systems. In: Koblitz, N. (ed.) CRYPTO 1996. LNCS, vol. 1109, pp. 104–113. Springer, Heidelberg (1996)

14. Mangard, S., Oswald, E., Popp, T.: Power Analysis Attacks: Revealing the Secrets of Smart Cards. Advances in Information Security. Springer-Verlag New York Inc., Secaucus (2007)

15. Mavroeidis, D., Batina, L., van Laarhoven, T., Marchiori, E.: PCA, eigenvector localization and clustering for side-channel attacks on cryptographic hardware devices. In: Flach, P.A., De Bie, T., Cristianini, N. (eds.) ECML PKDD 2012, Part I. LNCS, vol. 7523, pp. 253–268. Springer, Heidelberg (2012)

16. Messerges, T.S., Dabbish, E.A., Sloan, R.H.: Power analysis attacks of modular exponentiation in smartcards. In: Koç, Ç.K., Paar, C. (eds.) CHES 1999. LNCS, vol. 1717, p. 144. Springer, Heidelberg (1999)

17. Peeters, E., Standaert, F.X., Quisquater, J.J.: Power and electromagnetic analysis: improved model, consequences and comparisons. Integr. VLSI J. 40(1), 52–60 (2007)

18. Perin, G., Imbert, L., Torres, L., Maurine, P.: Attacking randomized exponentiations using unsupervised learning. In: Prouff, E. (ed.) COSADE 2014. LNCS, vol. 8622, pp. 144–160. Springer, Heidelberg (2014)

19. Rousseeuw, P.J.: Silhouettes: a graphical aid to the interpretation and validation of cluster analysis. J. Comput. Appl. Math. 20, 53–65 (1987)

20. Sauvage, L., Guilley, S., Mathieu, Y.: Electromagnetic radiations of FPGAs: high spatial resolution cartography and attack on a cryptographic module. ACM Trans. Reconfigurable Technol. Syst. 2, 4:1–4:24 (2009)

21. Souissi, Y., Bhasin, S., Guilley, S., Nassar, M., Danger, J.-L.: Towards different flavors of combined side channel attacks. In: Dunkelman, O. (ed.) CT-RSA 2012. LNCS, vol. 7178, pp. 245–259. Springer, Heidelberg (2012)

22. Standaert, F.-X., Archambeau, C.: Using subspace-based template attacks to compare and combine power and electromagnetic information leakages. In: Oswald, E., Rohatgi, P. (eds.) CHES 2008. LNCS, vol. 5154, pp. 411–425. Springer, Heidelberg (2008)

23. Veyrat-Charvillon, N., Gérard, B., Renauld, M., Standaert, F.-X.: An optimal key enumeration algorithm and its application to side-channel attacks. In: Knudsen, L.R., Wu, H. (eds.) SAC 2012. LNCS, vol. 7707, pp. 390–406. Springer, Heidelberg (2013)

24. Veyrat-Charvillon, N., Gérard, B., Standaert, F.-X.: Security evaluations beyond computing power. In: Johansson, T., Nguyen, P.Q. (eds.) EUROCRYPT 2013. LNCS, vol. 7881, pp. 126–141. Springer, Heidelberg (2013)

25. Walter, C.D.: Sliding windows succumbs to big mac attack. In: Koç, Ç.K., Naccache, D., Paar, C. (eds.) CHES 2001. LNCS, vol. 2162, p. 286. Springer, Heidelberg (2001)
26. Witteman, M.F., van Woudenberg, J.G.J., Menarini, F.: Defeating RSA multiply-always and message blinding countermeasures. In: Kiayias, A. (ed.) CT-RSA 2011. LNCS, vol. 6558, pp. 77–88. Springer, Heidelberg (2011)
27. Wolpert, D.H., Macready, W.G.: No free lunch theorems for optimization. IEEE Trans. Evol. Comput. **1**(1), 67–82 (1997)

Template Attacks vs. Machine Learning Revisited (and the Curse of Dimensionality in Side-Channel Analysis)

Liran Lerman [1]([✉]), Romain Poussier[2], Gianluca Bontempi [1],
Olivier Markowitch [1], and François-Xavier Standaert [2]

[1] Département d'informatique, Université Libre de Bruxelles, Brussels, Belgium
llerman@ulb.ac.be
[2] ICTEAM/INGI, Université catholique de Louvain, Louvain-la-Neuve, Belgium

Abstract. Template attacks and machine learning are two popular approaches to profiled side-channel analysis. In this paper, we aim to contribute to the understanding of their respective strengths and weaknesses, with a particular focus on their curse of dimensionality. For this purpose, we take advantage of a well-controlled simulated experimental setting in order to put forward two important intuitions. First and from a theoretical point of view, the data complexity of template attacks is not sensitive to the dimension increase in side-channel traces given that their profiling is perfect. Second and from a practical point of view, concrete attacks are always affected by (estimation and assumption) errors during profiling. As these errors increase, machine learning gains interest compared to template attacks, especially when based on random forests.

1 Introduction

In a side-channel attack, an adversary targets a cryptographic device that emits a measurable leakage depending on the manipulated data and/or the executed operations. Typical examples of physical leakages include the power consumption [15], the processing time [14] and the electromagnetic emanation [9].

Evaluating the security level of cryptographic implementations is an important concern, e.g. for modern smart cards. In this respect, profiled attacks are useful tools, since they can be used to approach their worst-case security level [24]. Such attacks essentially work in two steps: first a leakage model is estimated during a so-called profiling phase, then the leakage model is exploited to extract key-dependent information in an online phase. Many different approaches to profiling have been introduced in the literature. Template Attacks (TA), e.g. based on a Gaussian assumption [4], are a typical example. The stochastic approach exploiting Linear Regression (LR) is a frequently considered alternative [22]. More recently, solutions relying on Machine Learning (ML) have also been investigated [2,11–13,16,17,19]. These previous works support the claim that ML-based attacks are effective and lead to successful key recoveries. This is natural since they essentially exploit the same discriminating criteria as TA

© Springer International Publishing Switzerland 2015
S. Mangard and A.Y. Poschmann (Eds.): COSADE 2015, LNCS 9064, pp. 20–33, 2015.
DOI: 10.1007/978-3-319-21476-4_2

and LR (i.e. a difference in the mean traces corresponding to different intermediate computations if an unprotected implementation is targeted – a difference
in higher-order statistical moments if the device is protected with masking).
By contrast, it remains unclear whether ML can lead to more efficient attacks,
either in terms of profiling or in terms of online key recovery. Previous publications conclude in one or the other direction, depending on the implementation
scenario considered, which is inherent to such experimental studies.

In this paper, we aim to complement these previous works with a more sys-
tematic investigation of the conditions under which ML-based attacks may outperform TA (or not)[1]. For this purpose, we start with the general intuition that
ML-based approaches are generally useful in order to deal with high-dimensional
data spaces. Following, our contributions are twofold. First, we tackle the (theoretical) question whether the addition of useless (i.e. non-informative) leakage
samples in leakage traces has an impact on their informativeness if a perfect
profiling phase is achieved. We show that the (mutual) information leakage estimated with a TA exploiting such a perfect model is independent of the number
of useless dimensions if the useless leakage samples are independent of the useful
ones. This implies that ML-based attacks cannot be more efficient than template attacks in the online phase if the profiling is sufficient. Second, we study
the practical counterpart of this question, and analyze the impact of imperfect
profiling on our conclusions. For this purpose, we rely on a simulated experimental setting, where the number of (informative and useless) dimensions is used
as a parameter. Using this setting, we evaluate the curse of dimensionality for
concrete TA and compare it with ML-based attacks exploiting Support Vector
Machines (SVM) and Random Forests (RF). That is, we considered SVM as
a popular tool in the field of side-channel analysis, and RF as an interesting
alternative (since its random feature selection makes its behavior quite different than TA and SVM). Our experiments essentially conclude that TA outperform ML-based attacks whenever the number of dimensions can be kept reasonably low, e.g. thanks to a selection of Points of Interests (POI), and that ML
(and RF in particular) become(s) interesting in "extreme" profiling conditions
(i.e. with large traces and a small profiling sets) – which possibly arise when
little information about the target device is available to the adversary.

As a side remark, we also observe that most current ML-based attacks rate key
candidates according to (heuristic) scores rather than probabilities. This prevents
the computation of probability-based metrics (such as the mutual/perceived information [20]). It may also have an impact on the efficiency of key enumeration [25],
which is an interesting scope for further investigation.

The rest of the paper is organized as follows. Section 2 contains notations, the
attacks considered, our experimental setting and evaluation metrics. Section 3
presents our theoretical result on the impact of non-informative leakage samples in
perfect profiling conditions. Section 4 discusses practical (simulated) experiments

---

[1] Note that the gain of LR-based attacks over TA is known and has been analyzed,
e.g. in [10,23]. Namely, it essentially depends on the size of the basis used in LR.

in imperfect profiling conditions, in different contexts. Eventually, Sect. 5 concludes the paper and discusses perspectives of future work.

2   Background

2.1   Notations

We use capital letters for random variables and small caps for their realizations. We use sans serif font for functions (e.g. F) and calligraphic fonts for sets (e.g. $\mathcal{A}$). We denote the conditional probability of a random variable A given B with $\Pr[A|B]$ and use the acronym SNR for the signal-to-noise ratio.

2.2   Template Attacks

Let $l_{x,k}$ be a leakage trace measured on a cryptographic device that manipulates a target intermediate value $v = \mathsf{f}(x,k)$ associated to a known plaintext (byte) x and a secret key (byte) k. In a TA, the adversary first uses a set of profiling traces $\mathcal{L}_p$ in order to estimate a leakage model, next denoted as $\hat{\Pr}_{\mathrm{model}}\left[l_{x,k} \mid \hat{\theta}_{x,k}\right]$, where $\hat{\theta}_{x,k}$ represents the (estimated) parameters of the leakage Probability Density Function (PDF). The set of profiling traces is typically obtained by measuring a device that is similar to the target, yet under control of the adversary. Next, during the online phase, the adversary uses a set of new attack traces $\mathcal{L}_a$ (obtained by measuring the target device) and selects the secret key (byte) $\tilde{k}$ maximizing the product of posterior probabilities:

$$\tilde{k} = \underset{k^*}{\mathrm{argmax}} \prod_{l_{x,k} \in \mathcal{L}_a} \frac{\hat{\Pr}_{\mathrm{model}}\left[l_{x,k} \mid \hat{\theta}_{x,k^*}\right] \cdot \Pr[k^*]}{\hat{\Pr}_{\mathrm{model}}[l_{x,k}]}. \tag{1}$$

Concretely, the seminal TA paper suggested to use Gaussian estimations for the leakage PDF [4]. We will follow a similar approach and consider a Gaussian (simulated) experimental setting. It implies that the parameters $\hat{\theta}_{x,k}$ correspond to mean vectors $\hat{\mu}_{x,k}$ and covariance matrices $\hat{\Sigma}_{x,k}$. However, we note that any other PDF estimation could be considered by the adversary/evaluator [8]. We will further consider two types of TA: in the Naive Template Attack (NTA), we will indeed estimate one covariance matrix per intermediate value; in the Efficient Template Attack (ETA), we will pool the covariance estimates (assumed to be equal) across all intermediate values, as previously suggested in [5].

In the following, we will keep the $l_{x,k}$ and v notations for leakage traces and intermediate values, and sometimes omit the subscripts for simplicity.

2.3   Support Vector Machines

In their basic (two-classes) context, SVM essentially aims at estimating Boolean functions [6]. For this purpose, it first performs a supervised learning with labels (e.g. $v = -1$ or $v = 1$), annotating each sample of the profiling set. The binary SVM estimates a hyperplane $y = \hat{w}^\top l + \hat{b}$ that separates the two classes with the largest possible margin, in the geometrical space of the vectors. Then in the attack phase, any new trace l will be assigned a label $\tilde{v}$ as follows:

$$
\tilde{v} = \begin{cases} 1 & (\hat{w}^\top l + \hat{b}) \geq 1, \\ -1 & \text{otherwise.} \end{cases} \tag{2}
$$

Mathematically, SVM finds the parameters $\hat{w} \in \mathbb{R}^{n_s}$ (where n_s is the number of time samples per trace) and $\hat{b} \in \mathbb{R}$ by solving the convex optimization problem:

$$
\begin{aligned}
\min_{w,b} \quad & \tfrac{1}{2}(w^\top w), \\
\text{subject to} \quad & v(w^\top \phi(l_v) + b) \geq 1,
\end{aligned} \tag{3}
$$

where ϕ denotes a projection function that maps the data into a higher (sometimes infinite) dimensional space usually denoted as the feature space. Our experiments considered a Radial Basis kernel Function ϕ (RBF), which is a commonly encountered solution, both in the machine learning field and the side-channel communities. The RBF kernel maps the traces into an infinite dimensional Hilbert space in order to find a hyperplane that efficiently discriminate the traces. It is defined by a parameter γ that essentially relates to the "variance" of the model. Roughly, the variance of a model is a measure on the variance of its output in function of the variance of the profiling set. The higher the value of γ, the lower the variance of the model is. Intuitively, the variance of a model therefore relates to its complexity (e.g. the higher the number of points per trace, the higher the variance of the model). We always selected the value of γ as one over the number of points per trace, which is a natural choice to compensate the increase of the model variance due to the increase of the number of points per trace. Future works could focus on other strategies to select this parameter, although we do not expect them to have a strong impact on our conclusions.

When the problem of Eq. 3 is feasible with respect to the constraints, the data is said to be linearly separable in the feature space. As the problem is convex, there is a guarantee to find a unique global minimum. SVM can be generalized to multi-class problems (which will be useful in our context with typically 256 target intermediate values) and produce scores for intermediate values based on the distance to the hyperplane. In our experiments, we considered the *"one-against-one"* approach. In a one-against-one strategy, the adversary builds one SVM for each possible pair of target values. During the attack phase, the adversary selects the target value with a majority vote among the set of SVMs. Because of place constraints, we refer to [7] for a complete explanation.

2.4 Random Forests

Decision trees are classification models that use a set of binary rules to calculate a target value. They are structured as diagrams made of nodes and directed edges, where nodes can be of three types: root (i.e. the top node in the tree), internal (represented by a circle in Fig. 1) and leaf (represented by a square in Fig. 1). In our side-channel context, we typically consider decision trees in which (1) the value associated to a leaf is a class label corresponding to the target to be recovered, (2) each edge is associated to a test on the value of a time sample in the leakage traces, and (3) each internal node has one incoming edge from a node called the parent node, as also represented in Fig. 1.

In the profiling phase, learning data is used to build the model. For this purpose, the learning set is first associated to the root. Then, this set is split based on a time sample that most effectively discriminates the sets of traces associated to different target intermediate values. Each subset newly created is associated with a child node. The tree generator repeats this process on each derived subset in a recursive manner, until the child node contains traces associated to the same target value or the gain to split the subset is less than some threshold. That is, it essentially determines at which time sample to split, the value of the split, and the decision to stop or to split again. It then assigns terminal nodes to a class (i.e. intermediate value). Next, in the attack phase, the model simply predicts the target intermediate value by applying the classification rules to the new traces to classify. We refer to [21] for more details on decision trees.

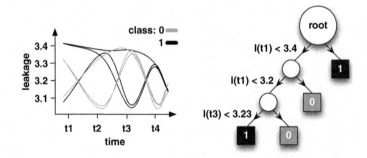

Fig. 1. Decision tree with two classes ($l(t1)$ is the leakage at time $t1$).

The Random Forests (RF) introduced by Breiman can be seen as a collection of classifiers using many (unbiased) decision trees as models [3]. It relies on model averaging (aka bagging) that leads to have a low variance of the resulting model. After the profiling phase, RF returns the most consensual prediction for a target value through a majority vote among the set of trees. RF are based on three main principles. First, each tree is constructed with a different learning set by re-sampling (with replacement) the original dataset. Secondly, the nodes of the trees are split using the best time sample among a subset of randomly chosen ones (by contrast to conventional trees where all the time samples are used). The size of this subset was set to the square of the number of time samples (i.e. $\sqrt{n_s}$) as suggested by Breiman. These features allow obtaining decorrelated trees, which improves the accuracy of the resulting RF model. Finally, and unlike conventional decision trees as well, the trees of a RF are fully grown and are not pruned, which possibly leads to overfitting (i.e. each tree has a low bias but a high variance) that is reduced by averaging the trees. The main (meta-) parameters of a RF are the number of trees. Intuitively, increasing the number of trees reduces the instability (aka variance) of the models. We set this number to 500 by default, which was sufficient in our experiments in order to show the strength of this model compared to template attack. We leave the detailed investigation of these parameters as an interesting scope for further research.

2.5   Experimental Setting

Let $l_{p,k}(t)$ be the t-th time sample of the leakage trace $l_{p,k}$. We consider contexts where each trace $l_{p,k}$ represents a vector of n_s samples, that is:

$$l_{p,k} = \{l_{p,k}(t) \in \mathbb{R} \mid t \in [1; n_s]\}. \tag{4}$$

Each sample represents the output of a leakage function. The adversary has access to a profiling set of N_p traces per target intermediate value, in which each trace has d informative samples and u uninformative samples (with $d + u = n_s$). The informative samples are defined as the sum of a deterministic part representing the useful signal (denoted as δ) and a random Gaussian part representing the noise (denoted as ϵ), that is:

$$l_{p,k}(t) = \delta_t(p, k) + \epsilon_t, \tag{5}$$

where the noise is independent and identically distributed for all t's. In our experiments, the deterministic part δ corresponds to the output of the AES S-box, iterated for each time sample and sent through a function G, that is:

$$\delta_t(p, k) = \mathsf{G}\left(\mathsf{SBox}^t(p \oplus k)\right), \tag{6}$$

where:

$$\mathsf{SBox}^1(p \oplus k) = \mathsf{SBox}(p \oplus k),$$
$$\mathsf{SBox}^t(p \oplus k) = \mathsf{SBox}\left(\mathsf{SBox}^{t-1}(p \oplus k)\right).$$

Concretely, we considered a function G that is a weighted sum of the S-box output bits. However, all our results can be generalized to other functions (preliminary experiments did not exhibit any deviation with highly non-linear leakage functions – which is expected in a first-order setting where the leakage informativeness essentially depends on the SNR [18]). We set our signal variance to 1 and used Gaussian distributed noise variables ϵ_t with mean 0 and variance σ^2 (i.e. the SNR was set to $\frac{1}{\sigma^2}$). Eventually, uninformative samples were simply generated with only a noisy part. This simulated setting is represented in Fig. 2 and its main parameters can be summarized as follows:

- Number of informative points per trace (denoted as d),
- Number of uninformative points per trace (denoted as u),
- Number of profiling traces per intermediate value (denoted as N_p),
- Number of traces in the attack step (noted N_a),
- Noise variance (denoted as σ^2) and SNR.

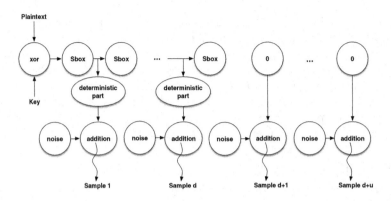

Fig. 2. Simulated leaking implementations.

2.6 Evaluation Metrics

The efficiency of side-channel attacks can be quantified according to various metrics. We will use information theoretic and security metrics advocated in [24].

Success Rate (SR). For an attack targeting a subkey (e.g. a key byte) and allowing to sort the different candidates, we define the success rate of order o as the probability that the correct subkey is ranked among the first o candidates. The success rate is generally computed in function of the number of attack traces N_a (given a model that has been profiled using N_p traces). In the rest of this paper, we focus on the success rate of order 1 (i.e. the correct key rated first).

Perceived/Mutual Information (PI/MI). Let X, K, L be random variables representing a target key byte, a known plaintext and a leakage trace. The perceived information between the key and the leakage is defined as [20]:

$$\hat{\mathrm{PI}}(K; X, L) = \mathrm{H}(K) + \sum_{k \in \mathcal{K}} \Pr[k] \sum_{x \in \mathcal{X}} \Pr[x] \sum_{l \in \mathcal{L}} \Pr_{\mathrm{chip}}[l|x, k] \cdot \log_2 \hat{\Pr}_{\mathrm{model}}[k|x, l].$$

The PI measures the adversary's ability to interpret measurements coming from the true (unknown) chip distribution $\Pr_{\mathrm{chip}}[l|x, k]$ with an estimated model $\hat{\Pr}_{\mathrm{model}}[l|x, k]$. $\Pr_{\mathrm{chip}}[l|x, k]$ is generally obtained by sampling the chip distribution (i.e. making measurement). Of particular interest for the next section will be the context of *perfect profiling*, where we assume that the adversary's model and the chip distribution are identical (which, strictly speaking, can only happen in simulated experimental settings since any profiling based on real traces will at least be imperfect because of small estimation errors [8]). In this context, the estimated PI will exactly correspond to the (worst-case) estimated MI.

Information theoretic metrics such as the MI/PI are especially interesting for the comparison of profiled side-channel attacks as we envision here. This is because they can generally be estimated based on a single plaintext (i.e. with $N_a = 1$)

whereas the success rate is generally estimated for varying N_a's. In other words, their scalar value provides a very similar intuition as the SR curves [23]. Unfortunately, the estimation of information theoretic metrics requires distinguishers providing probabilities, which is not the case of ML-based attacks[2]. As a result, our concrete experiments comparing TA, SVM and RF will be based on estimations of the success rate for a number of representative parameters.

3   Perfect Profiling

In this section, we study the impact of useless samples in leakage traces on the performances of TA with perfect profiling (i.e. the evaluator perfectly knows the leakages' PDF). In this context, we will use Pr for both Pr_{model} and Pr_{chip} (since they are equal) and omit subscripts for the leakages l to lighten notations.

Proposition 1. *Let us assume two TA with perfect models using two different attack traces l_1 and l_2 associated to the same plaintext x: l_1 is composed of d samples providing information and $l_2 = [l_1 || \epsilon]$ (where $\epsilon = [\epsilon_1, ..., \epsilon_u]$ represents noise variables independent of l_1 and the key.). Then the mutual information leakage $\text{MI}(K; X, L)$ estimated with their (perfect) leakage models is the same.*

Proof. As clear from the definitions in Sect. 2.6, the mutual/perceived information estimated thanks to TA only depend on $\text{Pr}[k|l]$. So we need to show that these conditional probabilities $\text{Pr}[k|l_2]$ and $\text{Pr}[k|l_1]$ are equal. Let k and k' represent two key guesses. Since ϵ is independent of l_1 and k, we have:

$$\frac{\text{Pr}[l_2|k']}{\text{Pr}[l_2|k]} = \frac{\text{Pr}[l_1|k'] \cdot \text{Pr}[\epsilon|k']}{\text{Pr}[l_1|k] \cdot \text{Pr}[\epsilon|k]},$$

$$= \frac{\text{Pr}[l_1|k'] \cdot \text{Pr}[\epsilon]}{\text{Pr}[l_1|k] \cdot \text{Pr}[\epsilon]},$$

$$= \frac{\text{Pr}[l_1|k']}{\text{Pr}[l_1|k]}. \tag{7}$$

This directly leads to:

$$\frac{\sum_{k' \in \mathcal{K}} \text{Pr}[l_2|k']}{\text{Pr}[l_2|k]} = \frac{\sum_{k' \in \mathcal{K}} \text{Pr}[l_1|k']}{\text{Pr}[l_1|k]},$$

$$\frac{\text{Pr}[l_2|k]}{\sum_{k' \in \mathcal{K}} \text{Pr}[l_2|k']} = \frac{\text{Pr}[l_1|k]}{\sum_{k' \in \mathcal{K}} \text{Pr}[l_1|k']},$$

$$\text{Pr}[k|l_2] = \text{Pr}[k|l_1], \tag{8}$$

which concludes the proof.

---

[2] There are indeed variants of SVM and RF that aim to remedy to this issue. Yet, the "probability-like" scores they output are not directly exploitable in the estimation of information theoretic metrics either. For example, we could exhibit examples where probability-like scores of one do not correspond to a success rate of one.

Quite naturally, this proof does not hold as soon as there are dependencies between the d first samples in l_1 and the u latter ones. This would typically happen in contexts where the noise at different time samples is correlated (which could then be exploited to improve the attack). Intuitively, this simple result suggests that in case of perfect profiling, the detection of POI is not necessary for a TA, since useless points will not have any impact on the attack's success. Since TA are optimal from an information theoretic point-of-view, it also means that the ML-based approaches cannot be more efficient in this context.

Note that the main reason why we need a perfect model for the result to hold is that we need the independence between the informative and non-informative samples to be reflected in these models as well. For example, in the case of Gaussian templates, we need the covariance terms that corresponds to the correlation between informative and non-informative samples to be null (which will not happen for imperfectly estimated templates). In fact, the result would also hold for imperfect models, as long as these imperfections do not suggest significant correlation between these informative and non-informative samples. But of course, we could not state that TA necessarily perform better than ML-based attacks in this case. Overall, this conclusion naturally suggests a more pragmatic question. Namely, perfect profiling never occurs in practice. So how does this theoretical intuition regarding the curse of dimensionality for TA extend to concrete profiled attack (with bounded profiling phases)? We study it in the next section.

4   Experiments with Imperfect Profiling

We now consider examples of TA, SVM- and RF-based attacks in order to gain intuition about their behavior in concrete profiling conditions. As detailed in Sect. 2, we will use a simulated experimental setting with various number of informative and uninformative samples in the leakage traces for this purpose.

4.1   Nearly Perfect Profiling

As a first experiment, we considered the case where the profiling is "sufficient" – which should essentially confirm the result of Proposition 1. For this purpose, we analyzed simulated leakage traces with 2 informative points (i.e. $d = 2$), $u = 0$ and $u = 15$ useless samples, and a SNR of 1, in function of the number of traces per intermediate value in the profiling set N_p. As illustrated in Fig. 3, we indeed observe that (e.g.) the PI is independent of u if the number of traces in the profiling set is "sufficient" (i.e. all attacks converge towards the same PI in this case). By contrast, we notice that this "sufficient" number depends on u (i.e. the more useless samples, the larger N_p needs to be). Besides, we also observe that the impact of increasing u is stronger for NTA than ETA, since the first one has to deal with a more complex estimation. Indeed, the ETA has 256 times more traces than the NTA to estimate the covariance matrice. So overall, and as expected, as long a the profiling set is large enough and the assumptions used to build the model capture the leakage samples sufficiently accurately, TA are indeed optimal, independent of the number of samples they actually profile. So there is little gain to expect from ML-based approaches in this context.

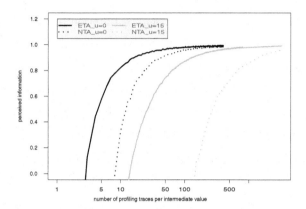

Fig. 3. Perceived information for NTA and ETA in function of N_p with SNR=1.

4.2 Imperfect Profiling

We now move to the more concrete case were profiling is imperfect. In our simulated setting, imperfections naturally arise from limited profiling (i.e. estimation errors): we will investigate their impact next and believe they are sufficient to put forward some useful intuitions regarding the curse of dimensionality in (profiled) side-channel attacks. Yet, we note that in general, assumption errors can also lead to imperfect models, that are more difficult to deal with (see, e.g. [8]) and are certainly worth further investigations. Besides, and as already mentioned, since we now want to compare TA, SVM and RF, we need to evaluate and compare them with security metrics (since the two latter ones do not output the probabilities required to estimate information theoretic metrics).

In our first experiment, we set again the number of useful dimensions to $d = 2$ and evaluated the success rate of the different attacks in function of the number of non-informative samples in the leakages traces (i.e. u), for different sizes of the profiling set. As illustrated in Fig. 4, we indeed observe that for a sufficient profiling, ETA is the most efficient solution. Yet, it is also worth observing that NTA provides the worst results overall, which already suggests that comparisons are quite sensitive to the adversary/evaluator's assumptions. Quite surprisingly, our experimental results show that up to a certain level, the success rate of RF increases with the number of points without information. The reason is intrinsic to the RF algorithm in which the trees need to be as decorrelated as possible. As a result, increasing the number of points in the leakage traces leads to a better independence between trees and improves the success rate. Besides, the most interesting observation relates to RF in high dimensionality, which remarkably resists the addition of useless samples (compared to SVM and TA). The main reason for this behavior is the random feature selection embedded into this tool. That is, for a sufficient number of trees, RF eventually detects the informative POI in the traces, which makes it less sensitive to the increase of u. By contrast, TA and SVM face a more and more difficult estimation problem in this case.

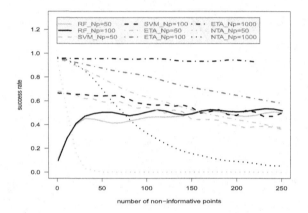

Fig. 4. Success rate for NTA, ETA, SVM and RF in fct. of the number of useless samples u, for various sizes of the profiling set N_p, with $d = 2$, SNR=1, $N_a = 15$.

Another noticeable element of Fig. 4 is that SVM and RF seem to be bounded to lower success rates than TA. But this is mainly an artifact of using the success rate as evaluation metric. As illustrated in Fig. 5 increasing either the number of informative dimensions in the traces d or the number of attack traces N_a leads to improved success rates for the ML-based approaches as well. For the rest, the latter figure does not bring significantly new elements. We essentially notice that RF becomes interesting over ETA for very large number of useless dimensions and that ETA is most efficient otherwise.

Eventually, the interest of the random feature selection in RF-based models raises the question of the time complexity for these different attacks. That is, such a random feature selection essentially works because there is a large enough number of trees in our RF models. But increasing this number naturally increases the time complexity of the attacks. For this purpose, we report some results regarding the time complexity of our attacks in Fig. 6. As a preliminary note, we mention that those results are based on prototype implementations in different programming languages (C for TA, R for SVM and RF). So they should only be taken as a rough indication. Essentially, we observe an overhead for the time complexity of ML-based attacks, which vanishes as the size of the leakage traces increases. Yet, and most importantly, this overhead remains comparable for SVM and RF in our experiments (mainly due to the fact that the number of trees was set to a constant 500). So despite the computational cost of these attacks is not

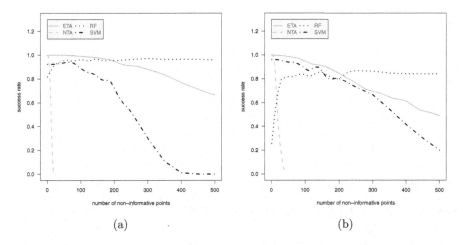

(a)   (b)

Fig. 5. (a) Success rate for NTA, ETA, SVM and RF in function of the number of useless samples u, with parameters $N_p = 25$, $d = 5$, SNR=1 and $N_a = 15$. (b) Similar experiment with parameters $N_p = 50$ $d = 2$, SNR=1 and $N_a = 30$.

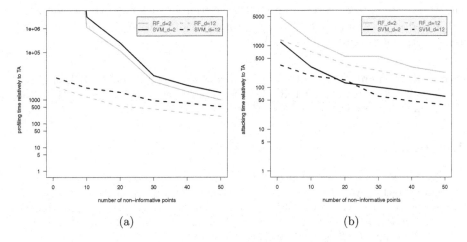

(a)   (b)

Fig. 6. Time complexity for ETA, SVM and RF in fct. of the number of useless samples, for $d = [2, 12]$ and $N_p = 25$. (a) Profiling phase. (b) Attack phase.

negligible, it remains tractable for the experimental parameters we considered (and could certainly be optimized in future works).

5 Conclusion

Our results provide interesting insights on the curse of dimensionality for side-channel attacks. From a theoretical point of view, we first showed that as long as a limited number of POI can be identified in leakage traces and contain most of the information, TA are the method of choice. Such a conclusion extends to any scenario where the profiling can be considered as "nearly perfect". By contrast, we also observed that as the number of useless samples in leakage traces increases and/or the size of the profiling set becomes too limited, ML-based attacks gain interest. In our simulated setting, the most interesting gain is exhibited for RF-based models, thanks to their random feature selection. Interestingly, the recent work of Banciu *et al.* reached a similar conclusion in a different context, namely, Simple Power Analysis and Algebraic Side-Channel Analysis [1].

Besides, and admittedly, the simulated setting we investigated is probably most favorable to TA, since only estimation errors can decrease the accuracy of the adversary/evaluator models in this case. One can reasonably expect that real devices with harder to model noise distributions would improve the interest of SVM compared to ETA – as has been suggested in previously published works. As a result, the extension of our experiments towards other distributions is an interesting avenue for further research. In particular, the study of leakage traces with correlated noise could be worth additional investigations in this respect. Meanwhile, we conclude with the interesting intuition that TA are most effi-cient for well understood devices, with sufficient profiling, as they can approach the worst-case security level of an implementation in such context. By contrast, ML-based attacks (especially RF) are promising alternative(s) in black box set-tings, with only limited understanding of the target implementation.

Acknowledgements. F.-X. Standaert is a research associate of the Belgian Fund for Scientific Research (FNRS-F.R.S.). This work has been funded in parts by the European Commission through the ERC project 280141 (CRASH).

References

1. Banciu, V., Oswald, E., Whitnall, C.: Reliable information extraction for single trace attacks. IACR Cryptology ePrint Archive, 2015:45 (2015)
2. Bartkewitz, T., Lemke-Rust, K.: Efficient template attacks based on probabilistic multi-class support vector machines. In: Mangard, S. (ed.) CARDIS 2012. LNCS, vol. 7771, pp. 263–276. Springer, Heidelberg (2013)
3. Breiman, L.: Random forests. Mach. Learn. **45**(1), 5–32 (2001)
4. Chari, S., Rao, J.R., Rohatgi, P.: Template attacks. In: Kaliski Jr., B.S., Koç, Ç.K., Paar, C. (eds.) CHES 2002. LNCS, vol. 2523, pp. 13–28. Springer, Heidelberg (2002)
5. Choudary, O., Kuhn, M.G.: Efficient template attacks. In: Francillon, A., Rohatgi, P. (eds.) CARDIS 2013. LNCS, vol. 8419, pp. 253–270. Springer, Heidelberg (2014)
6. Cortes, C., Vapnik, V.: Support-vector networks. Mach. Learn. **20**(3), 273–297 (1995)
7. Cristianini, N., Shawe-Taylor, J.: An Introduction to Support Vector Machines and Other Kernel-based Learning Methods. Cambridge University Press, Cambridge (2010)
8. Durvaux, F., Standaert, F.-X., Veyrat-Charvillon, N.: How to certify the leak-age of a chip? In: Nguyen, P.Q., Oswald, E. (eds.) EUROCRYPT 2014. LNCS, vol. 8441, pp. 459–476. Springer, Heidelberg (2014)

9. Gandolfi, K., Mourtel, C., Olivier, F.: Electromagnetic analysis: concrete results. In: Koç, Ç.K., Naccache, D., Paar, C. (eds.) CHES 2001. LNCS, vol. 2162, pp. 251–261. Springer, Heidelberg (2001)

10. Gierlichs, B., Lemke-Rust, K., Paar, C.: Templates vs. stochastic methods. In: Goubin, L., Matsui, M. (eds.) CHES 2006. LNCS, vol. 4249, pp. 15–29. Springer, Heidelberg (2006)

11. Heuser, A., Zohner, M.: Intelligent machine homicide. In: Schindler, W., Huss, S.A. (eds.) COSADE 2012. LNCS, vol. 7275, pp. 249–264. Springer, Heidelberg (2012)

12. Hospodar, G., Gierlichs, B., De Mulder, E., Verbauwhede, I., Vandewalle, J.: Machine learning in side-channel analysis: a first study. J. Cryptographic Eng. **1**(4), 293–302 (2011)

13. Hospodar, G., De Mulder, E., Gierlichs, B., Vandewalle, J., Verbauwhede, I.: Least squares support vector machines for side-channel analysis. In: Second International Workshop on Constructive Side-Channel Analysis and Secure Design, pp. 99–104. Center for Advanced Security Research Darmstadt (2011)

14. Kocher, P.C.: Timing attacks on implementations of Diffie-Hellman, RSA, DSS, and other systems. In: Koblitz, N. (ed.) CRYPTO 1996. LNCS, vol. 1109, pp. 104–113. Springer, Heidelberg (1996)

15. Kocher, P.C., Jaffe, J., Jun, B.: Differential power analysis. In: Wiener, M. (ed.) CRYPTO 1999. LNCS, vol. 1666, pp. 388–397. Springer, Heidelberg (1999)

16. Lerman, L., Bontempi, G., Markowitch, O.: Side-channel attacks: an approach based on machine learning. In: Second International Workshop on Constructive Side-Channel Analysis and Secure Design, pp. 29–41. Center for Advanced Security Research Darmstadt (2011)

17. Lerman, L., Bontempi, G., Markowitch, O.: Power analysis attack: an approach based on machine learning. IJACT **3**(2), 97–115 (2014)

18. Mangard, S., Oswald, E., Standaert, F.-X.: One for all - all for one: unifying standard differential power analysis attacks. IET Inf. Secur. **5**(2), 100–110 (2011)

19. Patel, H., Baldwin, R.O.: Random forest profiling attack on advanced encryption standard. IJACT **3**(2), 181–194 (2014)

20. Renauld, M., Standaert, F.-X., Veyrat-Charvillon, N., Kamel, D., Flandre, D.: A formal study of power variability issues and side-channel attacks for nanoscale devices. In: Paterson, K.G. (ed.) EUROCRYPT 2011. LNCS, vol. 6632, pp. 109–128. Springer, Heidelberg (2011)

21. Rokach, L., Maimon, O.: Data Mining with Decision Trees: Theory and Applications. Series in machine perception and artificial intelligence. World Scientific Publishing Company, Incorporated, Singapore (2008)

22. Schindler, W., Lemke, K., Paar, C.: A stochastic model for differential side channel cryptanalysis. In: Rao, J.R., Sunar, B. (eds.) CHES 2005. LNCS, vol. 3659, pp. 30–46. Springer, Heidelberg (2005)

23. Standaert, F.-X., Koeune, F., Schindler, W.: How to compare profiled side-channel attacks? In: Abdalla, M., Pointcheval, D., Fouque, P.-A., Vergnaud, D. (eds.) ACNS 2009. LNCS, vol. 5536, pp. 485–498. Springer, Heidelberg (2009)

24. Standaert, F.-X., Malkin, T.G., Yung, M.: A unified framework for the analysis of side-channel key recovery attacks. In: Joux, A. (ed.) EUROCRYPT 2009. LNCS, vol. 5479, pp. 443–461. Springer, Heidelberg (2009)

25. Veyrat-Charvillon, N., Gérard, B., Renauld, M., Standaert, F.-X.: An optimal key enumeration algorithm and its application to side-channel attacks. In: Knudsen, L.R., Wu, H. (eds.) SAC 2012. LNCS, vol. 7707, pp. 390–406. Springer, Heidelberg (2013)

Efficient Selection of Time Samples
for Higher-Order DPA with Projection Pursuits

François Durvaux[1]([✉]), François-Xavier Standaert[1], Nicolas Veyrat-Charvillon[2],
Jean-Baptiste Mairy[3], and Yves Deville[3]

[1] ICTEAM/ELEN/Crypto Group, Université catholique de Louvain,
Louvain-la-Neuve, Belgium
francois.durvaux@uclouvain.be
[2] IRISA-CAIRN, Campus ENSSAT, 22305 Lannion, France
[3] ICTEAM/INGI, Université catholique de Louvain, Louvain-la-Neuve, Belgium

Abstract. The selection of points-of-interest in leakage traces is a frequently neglected problem in the side-channel literature. However, it can become the bottleneck of practical adversaries/evaluators as the size of the measurement traces increases, especially in the challenging context of masked implementations, where only a combination of multiple shares reveals information in higher-order statistical moments. In this paper, we describe new (black box) tools for efficiently dealing with this problem. The proposed techniques exploit projection pursuits and specialized local search algorithms, work with minimum memory requirements and practical time complexity. We validate them with two case-studies of unprotected and first-order masked implementations in an 8-bit device, the latter one being hard to analyze with previously known methods.

1 Introduction

The selection of Points-Of-Interest (POIs) in leakage traces is an important (and not very discussed) problem in the application of Side-Channel Analysis (SCA) attacks. When targeting unprotected implementations, the naive strategy that is commonly used in the literature is to test all the time samples independently. It raises two important challenges. First, how to combine these time samples efficiently, in order to maximize the amount of information extracted from each leakage trace? Second, how to extend this technique in the context of masked implementations where the sensitive data is split into d shares manipulated in different clock cycles (as it is typically the case in software), and only the combination of these shares' leakage reveals key-dependent information – which makes the complexity of an exhaustive analysis grow combinatorially with d?

Solutions to the first problem typically include dimensionality reduction techniques such as PCA and LDA. These tools (introduced to SCA in [1,23] and recently revisited in [3,5]) essentially project the leakage traces into a lower-dimensional subspace that optimizes some *objective function*. Namely, PCA usually maximizes the variance between the mean leakage traces – i.e. the signal of

© Springer International Publishing Switzerland 2015
S. Mangard and A.Y. Poschmann (Eds.): COSADE 2015, LNCS 9064, pp. 34–50, 2015.
DOI: 10.1007/978-3-319-21476-4_3

a first-order DPA, while LDA maximizes the ratio between inter-class and intra-class variances – i.e. its Signal-to-Noise Ratio (SNR), essentially. Their main advantage is to provide a principled and intuitive solution to the problem, since the projection (i.e. eigenvectors) they produce indicate the POIs. Yet, they are somewhat limited when moving to masked implementations for which the information lies in high-order statistical moments, since their objective function is based on a definition of signal that primarily captures first-order leakages[1].

Solutions to the second problem are even sparser. To the best of our knowledge, the usual reference for selecting POIs for masked implementations is the educated guess proposed by Oswald et al. in [13] (i.e. an exhaustive search over all d-tuples of time samples in a window selected based on engineering intuition). Next, Reparaz et al. proposed an alternative solution exploiting Mutual Information Analysis (MIA) [8], that allows gaining a constant (but practically meaningful) factor corresponding the number of key hypotheses in the attack [20]. In both cases, the proposed tools do not output a projection but a list of the most useful POIs (i.e. d-tuples) in function of the (non-profiled) attack considered.

In this paper, we investigate the use of *Projection Pursuits* (PPs), as alternative tools for the selection of POIs in leakage traces [7]. Intuitively, PPs machine-pick "interesting" low-dimensional projections of a high-dimensional data space by numerically maximizing a certain objective function. They essentially work by tracking the improvements (or lack thereof) of the projection when modifying it with small random perturbations. Their main advantage in our context is that they can deal with any objective function, which naturally fits to the problem of higher-order SCA. Their main drawback is (in general) their heuristic nature, since the convergence of the method is not guaranteed and its complexity is context-dependent. As a result, and in order to validate the interest of PPs in our SCA context, we first applied them to the simple case of an unprotected implementation of the AES. We show that different objective functions can be efficiently used for this purpose, leading to powerful subspace-based attacks, with similar informativeness as previous solutions such as LDA.

Next, we moved to the more challenging context of masking. In this case, we combined the (linear) projection with an objective function exploiting higher-order statistical moments. Initial experiments suggest that the straightforward implementation of a PP algorithm is not efficient in detecting the POIs of such protected implementations (especially as the number of useless dimensions in the traces increases). The main reason is that as long as a d-tuple of POIs is not present in the projection, the objective function essentially returns random indications. Interestingly, we then show that a specialized PP algorithm exploiting an improved *local search* could give excellent results even in this challenging context. Intuitively, it works by looking for the best size and position of d windows covering parts of the traces, again by iterating small random perturbations. Our experiments suggest that we can recover POIs with significantly less calls to

---

[1] Of course, a trivial solution would be to apply PCA/LDA to "product traces" containing all the possible products of d-tuples, but this rapidly leads to unrealistic memory requirements in the masked software context that we consider next.

the objective function than a exhaustive analysis. We further discuss the main parameters influencing the success of such a detection method, and detail the time vs. measurement complexity tradeoff resulting from these parameters.

2   Background

Notations. We use capital letters for random variables, small caps for their realizations, sans serif fonts for functions and calligraphic letters for sets.

2.1   Measurement Setups

Our experiments are based on measurements of an AES implementation run by an 8-bit Atmel AVR (ATMega644P) microcontroller at a 20 MHz clock frequency. We monitored the voltage variations across a 22 Ω resistor introduced in the supply circuit of our target chip. Acquisitions were performed using a Lecroy HRO66ZI oscilloscope running at 200 MHz and providing 8-bit samples. For concreteness, our evaluations focused on the leakage of the first AES master key byte (but would apply identically to any other enumerable target). Leakage traces were produced according to the following procedure. Let x and s be our target input plaintext byte and subkey, and $y = x \oplus s$ denote a key addition. For each of the 256 values of y, we generated 1000 unprotected encryption traces (resp. 500 for masked traces), where the rest of the plaintext and key was random, i.e. we generated 256 000 (resp. 128 000) traces in total, with plaintexts of the shape $p = x||r_1||\dots||r_{15}$, keys of the shape $k = s||r_{16}||\dots||r_{30}$, and the r_i's denoting uniformly random bytes. In case of masked implementations, additional uniform randomness was used to generate the shares. In order to reduce the memory cost of our evaluations, we only stored the leakages corresponding to the 2 first AES rounds in the unprotected case (as the dependencies in our target byte $y = x \oplus s$ typically vanish after the first round, because of the strong diffusion properties of the AES). As for the protected case, we only considered a single S-box, for which the precomputation of a masked table alreay implies large traces with $Ns = 30,000$ time samples (vs. $Ns = 1500$ for the unprotected one). As will be clear next, these sets of measurements were large enough to emphasize the interest of our projection pursuit algorithms. In the following, we will denote the 1000 (resp. 500) encryption traces obtained from a plaintext p including the target byte x under a key k including the subkey s as: $\mathsf{AES}_{k_s}(p_x) \rightsquigarrow l_y^i$, with $i \in [1; 1000]$ (resp. $i \in [1; 500]$). Whenever accessing the points of these traces, we will additionally use an argument t (for time), leading to $l_y^i(t)$. Our goal is to generate projections exhibiting the time samples that contain information about y. Note that since we assume the plaintext to be known by the adversary (as usual in SCAs), it directly translates into information about s – which typically occurs during the key addition $y = x \oplus s$ and S-box execution $z = \mathsf{S}(x \oplus s)$.

2.2   Objective Functions (Aka Evaluation Metrics)

In order to "guide" the PP, we need to define criteria to determine whether some modification of the projection is positive. Any SCA evaluation metric can be used for this purpose. We list a few candidates in this section. In order to guarantee their soundness, we focused on objective functions based on profiled distinguishers (which allows mitigating biases due to incorrect a-priori choices of models – given that the profiles are well estimated and based on sound assumptions).

CPA [4]. In a profiled Correlation Power Analysis, the adversary first estimates the first-order moments corresponding to each value y from a vector of N_p profiling traces $\mathbf{l}_y^p$, that we denote as $\hat{\mathbf{m}}_y^1 = \hat{\mathsf{E}}(\mathbf{l}_y^p)$, with $\hat{\mathsf{E}}$ the sample mean operator. This step is performed for each time sample independently, leading to $\hat{\mathbf{m}}_y^1(t)$. Since there are 256 y values in our AES case study, it amounts to compute $256 \times N_s$ means, with N_s the number of samples per trace. Then, he computes the correlation between these mean values and the samples coming from a vector of test traces $\mathbf{l}_y^t$, leading to $\hat{\rho}(\hat{\mathbf{m}}_y^1(t), \mathbf{l}_y^t(t))$ with $\hat{\rho}$ denoting Pearson's coefficient.

SNR [10]. An alternative to CPA is the SNR defined at CT-RSA 2004 as:

$$\hat{\mathsf{SNR}}(t) = \frac{\hat{\mathsf{var}}_y\left(\hat{\mathsf{E}}(\mathbf{l}_y^t(t))\right)}{\hat{\mathsf{E}}_y\left(\hat{\mathsf{var}}(\mathbf{l}_y^t(t))\right)},$$

with $\hat{\mathsf{var}}$ the sample variance operator. Similarly to the correlation coefficient, such a criteria is discriminant for first-order information (i.e. information lying in the first-order moments of the leakage distribution). In order to deal with masked implementations, we also need objective functions that capture more general dependencies. In this context, a natural option is the information theoretic metric introduced in [25] and later refined in [19]. Its sample definition is given by:

$$\hat{\mathsf{I}}(S; X, L) = \mathsf{H}[S] + \sum_{s \in \mathcal{S}} \Pr[s] \sum_{x \in \mathcal{X}} \Pr[x] \sum_{l_y^i \in \mathcal{L}_Y^t} \Pr_{\mathsf{chip}}[l_y^i | s, x] . \log_2 \hat{\Pr}_{\mathsf{model}}[s | x, l_y^i],$$

where $\hat{\Pr}_{\mathsf{model}}$ is a probabilistic model estimated thanks to the set of profiling traces (just as the $256 \times N_s$ mean values in the correlation case). Computing such an objective function implies (constant but significant) performance overheads, because it requires applying Bayes' law and marginalizing over the key hypotheses. Since the objective function will typically be applied after projection in the following sections (i.e. in a univariate context), a cheaper alternative is to exploit the following "Moments-Correlating Profiled DPA" (MCP-DPA):

MCP-DPA [12]. The attack features essentially the same steps as a profiled CPA. The only difference is that the adversary will estimate dth-order moments $\hat{\mathbf{m}}_y^d(t)$ with the profiling traces. In the following, we will be particularly interested in the Moments against Moments Profiled Correlation (MMPC) criteria:

$$\mathrm{MMPC}(t) = \hat{\rho}(\hat{\mathbf{m}}_y^d(t), \tilde{\mathbf{m}}_y^d(t)),$$

where $\tilde{\mathbf{m}}_y^d(t)$ are another vector of moments, estimated with the test traces. As detailed in [12], MCP-DPA is able to capture information in any statistical moment, while enjoying the implementation efficiency of CPA (which is highly beneficial in our context where the objective function is intensively used).

3 Projection Pursuit Against Unprotected Devices

In this section we investigate the application of PPs to the simple case of the
(unprotected) AES furious implementation available as open source from [15].
In this context, our goal is to find a projection vector $\boldsymbol{\alpha}$ that will convert the N_s
samples of a leakage vector $\mathbf{l}_y^{\mathrm{t}}$ to a single (projected) sample λ_y^i, that is:

$$\lambda_y^i = \sum_{t=0}^{N_s-1} \alpha(t) \cdot l_y^i(t),$$

such that univariate attacks exploiting the λ_y^i's will be most efficient. This essentially requires to define an objective function that measures the "informativeness" of these samples. As mentioned in the previous section, this task is quite
easy when first-order information is available in the leakage traces: Pearson's
correlation coefficient obtained from a CPA and Mangard's SNR are natural
candidates – we will try them both in the next subsection. Following the equivalence results in [11], they should provide similar results in this case (also similar
to the ones that would be obtained with an information theoretic metric).

3.1 Projection Pursuit Algorithm

The pseudo-code of our projection pursuit algorithm is given in Algorithm 1.

Algorithm 1. Projection Pursuit.

---

basic_PP(N_r,N_{it})
$\boldsymbol{\alpha} = initialize()$;
repeat N_r times
    $r = rand\_index(N_s)$;
    $\boldsymbol{\alpha}_{new} = max\_search(@f_{obj}, \mathcal{L}^{\mathrm{p}}, \boldsymbol{\alpha}, r, N_{it})$;
    $\boldsymbol{\alpha} = \boldsymbol{\alpha}_{new}$;
end

It essentially repeats (N_r times) the selection of a random index r followed by a
maximization of the objective function for the corresponding time sample, based
on the set of profiling traces $\mathcal{L}^{\mathrm{p}}$ (which contains traces for all the intermediate
values y). For this purpose, the $max\_search()$ function consists in successive
parabolic interpolations (illustrated in the long version of this work [6]), which
work in two iterated steps. We first look for samples that enclose the extremum
as follows. From a starting point x_1, we add a Δ in the direction that increases
f_{obj} to get x_2. Then, we keep adding Δ's until finding x_3 such that $y_3 < y_2$.
As the weights assigned to each time sample are between 0 and 1, we typically
take Δ's corresponding to a couple of percents (e.g. 0.1 in our experiments) and
repeat such additions at most $1/\Delta$ times. Then, based on these three points,

we start interpolating. This process is iterated N_{it} times, during which we replace the "oldest" x-point by the x-coordinate (x_v) of the parabola vertex (y-values are re-computed accordingly). The new $\alpha(t)$ gets its value from the median x-value at the end of the last iteration. In our experiments, $N_{it} = 3$ iterations were enough to get a good approximation of the maximum. This method has the advantage of being very fast to compute and to converge. Note finally that the number of repetitions N_r should ideally be larger than the number of samples N_s (e.g. twice, typically), because some weights benefit from being re-adjusted after the modification of other $\alpha(t)$'s. Yet, when applied in the context of an unprotected implementation, the time complexity of Algorithm 1 was never a practical limitation in our experiments (i.e. a few minutes of computations).

3.2   Experimental Results

We implemented the PP algorithm for both the CPA and SNR objective functions, and targeted the first AES key byte for illustration. For each of the 256 values of $y = x \oplus s$, we measured $N_p = N_t = 50$ traces for the CPA objective function, and $N_t = 100$ traces for the SNR one, each of them made of $N_s = 1500$ time samples. We set N_r, N_{it} and Δ as just explained (to 3000, 3 and 0.1, respectively). We then computed success rates to compare the quality of the projections obtained with the most informative sample, by performing 2000 experimental univariate Template Attacks (TA). These results show the effectiveness of the projections as they need only 7 traces to get a 90 % success rate, against 28 traces for the univariate TA. It also confirms that both objective functions are indeed equivalent in this case. It is finally interesting to compare our findings with the results in [24] that target a similar implementation (with very similar success rate for the univariate TA). In particular, we see that the univariate attack based on the single sample provided by our projections leads to approximately the same data complexities as the hexavariate template attack taking (heuristic) advantage of all the POIs in this previous work (Fig. 1).

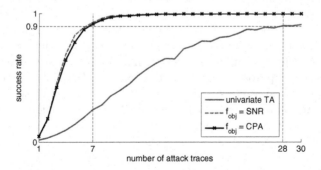

Fig. 1. Template attack success rates against unprotected device

4   Projection Pursuit Against Masked Implementations

In contrast with the previous section, detecting POIs in leakage traces of masked implementations is a quite challenging task. From the complexity point-of-view, exhaustive approaches may grow exponentially with the number of shares (if these shares are manipulated at different time samples), making them unpractical for long traces. Furthermore, the information in the leakages of masked implementation lies in higher-order moments of their probability distribution, which are harder to estimate. As a result, the direct application of Algorithm 1 with the previous objective functions in this context does not provide successful results. In the (simple) case where the shares of a masking scheme are manipulated in parallel, adapting the objective function may be sufficient to deal with this problem. But in case of software implementations, where the shares are manipulated at different time samples, it is the algorithm itself that has to be adapted. Intuitively, this is because it works by modifying time samples one at a time, while for such masked implementation, we require at least one meaningful d-tuple of samples to be active in the projection for an objective function to output relevant information. We now describe how to specialize PPs to take this constraint into account, and detect POIs for masked implementations.

4.1   Specialized Projection Pursuit Algorithm

The main tool used in our following optimization is local search, which is a collection of iterative methods that are efficient for quickly finding good solutions to optimization problems (note that the previous PP algorithm can be viewed as a simple local search). Despite heuristic, it generally works more efficiently than exhaustive analyses. Furthermore, local search has very limited storage requirements. For example, in our context, it exploits the leakage traces directly – which is a significant advantage compared to heuristics exploiting "product traces" as mentioned in footnote 1. A good reference to these methods is [9]. Their working principle is simple: they always keep a solution (called the current solution) as well as the best solution found since the beginning of the search. At each iteration of the algorithm, the current solution is perturbed, giving a set of new solutions, called its neighborhood. One of the neighboring solutions is then selected and replaces the current solution. The algorithm terminates when its convergence criterion is met (e.g. number of iterations without improvement, time limit, etc.). Intuitively, such an approach to optimization exploits *diversification* and *intensification*. The former aims at exploring a large and diverse search space, while the latter intends to improve the current solution. Their combination is expected to avoid being trapped into local optima.

When applied to masking, one key element has to be taken into account by optimizations. Namely, the sensitive variables are split into d shares and the objective function should not be informative as long as a meaningful d-tuple of shares is not present in the projection. Besides, in practice it frequently happens that dimensions near a POI also contain valuable information. These two facts motivate the way we designed our improved search algorithm as follows. First, we consider a projection vector containing d *windows* of non-zero weights (all the others being zero) and denote a group of successive dimensions as a window. The weights inside these windows are uniform. In this context, and since local search only considers local modifications of the current solution, the information given by the objective function will return essentially random indications (so no reliable information) if this current solution does not cover the d shares. On the contrary, when the windows spans a d-tuple of shares, the objective function can be used to refine the current solution. For this reason, our specialized PP algorithm will be split into two parts next denoted as find_sol and improve_sol. The find_sol phase probes the search space with large windows and a lot of randomness until it has good indication that the windows span the d-tuples of shares. In order to detect that the windows span these d-tuples, we use two sets of profiling traces ($\mathcal{L}_{tr}^{p}$ and $\mathcal{L}_{va}^{p}$, where tr stands for training and va for validation). Then, the improve_sol phase refines those windows. The find_sol phase thus puts more emphasis on diversification and the improve_sol, on intensification.

Algorithm 2. Specialized projection pursuit algorithm using local search.

specialized_PP_Local_Search(d, W_{len}, T_{det},TP:=TP'∪ TP'')
   α = find_sol_phase(d, W_{len}, T_{det},TP');
   if($\alpha \neq null$)
      return improve_sol_phase(α,TP'');
   end

The pseudocodes of the specialized PP algorithm using local search are given in Algorithms 2, 3 and 4. These algorithms depend on various parameters: some of them will be explicitly discussed as they hold important intuitions, the remaining ones – next denoted as technical parameters (TP) – will be fixed according to state-of-the-art strategies. Our main tool is the specialized_PP_Local_Search function (Algorithm 2). As just explained, it organizes the search in two main steps. The first one is the find_sol phase which returns a first candidate projection α (after N_r^f repetitions). If this first step is successful, the improve_sol phase is repeated N_r^i times to refine the solution. The find_sol phase is described in Algorithm 3. At each iteration, it randomly selects d windows of length W_{len} with non-zero weights (function random_window). All the neighbors of the solution are then computed with the function get_neighbors_FS. Each neighbor is constructed by moving one of the windows left or right (if we see the projection vector as a row vector). The lengths of the moves considered are small multiples of the window length (as set by the *num_hops* parameter). During the computation of the neighbors, the collisions between windows are avoided in order to keep d distinct windows. Next, the best neighbor is selected as the neighbor having the maximal evaluation of f_{obj} on the set $\mathcal{L}_{tr}^{p}$. This best neighbor is finally tested to detect if a d-tuple of shares is spanned by the windows. The detection is based on a threshold T_{det} on the objective function that will be carefully discussed in the next section. In order to dodge the randomness of the objective function when the d shares are not spanned, this threshold has to be exceeded on both the training and validation sets of traces $\mathcal{L}_{tr}^p$, $\mathcal{L}_{va}^p$. If those two conditions are met, the projection vector is returned by the algorithm.

Algorithm 3. Find solution phase.

---

find_sol_phase($d, W_{len}, T_{det},$ TP')
TP':=$\{N_r^f, num\_hops\}$
   i=0;
   repeat N_r^f times
      $\alpha = random\_window(d, W_{len})$;
      $neighborhood = get\_neighbors\_FS(\alpha, num\_hops)$;
      $best\_neighbor = max(@f_{obj}, neighborhood, \mathcal{L}_{tr}^p)$;
      if $f_{obj}(best\_neighbor, \mathcal{L}_{tr}^p) > T_{det}$ & $f_{obj}(best\_neighbor, \mathcal{L}_{va}^p) > T_{det}$
         return $(i + 1, best\_neighbor)$;
      end
      i++;
   end
end

If the find_sol phase was able to find a solution spanning the d shares, the objective function is informative enough to allow a second (intensification) step, and the improve_sol phase (in Algorithm 4) is run for N_r^i iterations. At each iteration, the entire neighborhood is constructed with the function get_neighbors_IS. Each neighbor results from the shift (left or right) of one window or the resizing of all the windows (we keep the same size for all windows). The move steps considered are given in *move_steps*, and the resize steps in *resize_steps*. The size of the windows is constrained to remain between *min_WS* and *max_WS*. The selection of the neighbor is then performed by select_neighbor, as a random neighbor amongst the N_n best neighbors. Using this selection strategy allows the search to avoid being trapped into local optima, ensuring a sufficient diversification. The search also memorizes the best projection obtained since the beginning of the phase in α_{best}. This is mandatory as it is allowed to select projection vectors that decrease the objective function. Eventually, the variable *num_stagn* records the number of iterations without improvement of the best solution. Once *num_stagn* is larger than *max_stagn* or when the number of iterations reaches $N_r = N_r^f + N_r^i$, the search returns the best solution α_{best}.

As far as the technical parameters are concerned, we first set the number of hops (*num_hops*) in the find_sol phase to allow the windows covering all the dimensions of the traces. It enables an iteration to find a covering set of windows when one window is incorrectly placed. Next, in the improve_sol phase, the more move steps (*move_steps*) and resize steps (*resize_steps*), the quicker the algorithm converges towards the optimal windows, but the longer each iteration is. We found that a good tradeoff in our context was to use *move_steps* of 1,

---

Algorithm 4. Improve solution phase.

---

improve_sol_phase(α,TP'')

TP'':=$\{N_r^i, move\_steps, resize\_steps, minWS, maxWS, N_n, max\_stagn\}$

   $\alpha_{best} = \alpha$;

   Repeat N_r^i times

      $neighborhood = get\_neighbors\_IS(\alpha, move\_steps, resize\_steps, minWS, maxWS)$;

      $\alpha = select\_neighbor(@f_{obj}, \mathcal{L}_{tr}^p, N_n)$;

      if $f_{obj}(\alpha, \mathcal{L}_{tr}^p) > f_{obj}(\alpha_{best}, \mathcal{L}_{tr}^p)$

         $\alpha_{best} = \alpha$;

         $num\_stagn = 0$;

      else

         $num\_stagn + +$;

      end

      if $num\_stagn > max\_stagn$

         return α_{best};

      end

   end

   return α_{best};

end

---

3 or 5 dimensions and *resize_steps* of 1 dimension. Those settings allow the iterations to be fast while still covering a large part of the search space around the solution found by the **find_sol** phase. The *min_WS* parameter typically depends on the sampling rate of the oscilloscope used in the attack: we set it to 5 which corresponds to half a cycle in our experiments, based on the intuition that dimensions next to a POI may also contain information. *max_WS* was then chosen as $2*W_{len}$, reflecting that this information can be spread on multiple clock cycles. Finally, a *max_stagn* value of 50 allows the local search to stop when it is unlikely to further improve the quality of the windows. And given the low span of the moves and the resizes, an exploration parameter N_n of 3 is enough to escape local optima and still converge towards the optimal solution.

4.2 Simulated Experiments

We now discuss the setting of the more intuitive parameters W_{len} and T_{det} together with the performance gains obtained thanks to our specialized PP algorithm. In view of their heuristic nature, these questions are best investigated with simulated examples, where we can play with some important parameters of leaking implementations. For this purpose, we will consider a first-order masked S-box where the adversary receives N_i pairs of leakage variables of the form:

$$L_i^1 = \mathsf{HW}(\mathsf{S}(x \oplus s) \oplus m) + R_i^1,$$
$$L_i^2 = \mathsf{HW}(m) + R_i^2, \tag{1}$$

where HW is the Hamming weight function, S the AES S-box, x a plaintext byte, s a key byte, m a secret random mask, and R_i^1, R_i^2 are normally distributed noise variables with variance σ_n^2 ($1 < i \leq N_i$). For simplicity, we make sure that the N_i samples corresponding to the two shares are not overlapping. Next to these $2 \times N_i$ informative samples, we finally add $N_s - 2 \times N_i$ random samples N_j, so that N_s is the total number of samples in our simulated traces.

Setting the Detection Threshold. An important parameter in Algorithm 3 is the threshold value used to decide whether an improvement of the objective function is significant. In this context, a particularly convenient feature of the MMPC criteria (defined in Sect. 2.2) is that it gradually tends to one as the number of measurements used in the detection increases. That is, given that the order of the statistical moment (e.g. $d = 2$ in our current simulations) and number of measurements used in the detection is sufficient, this criteria always reaches high values. Intuitively, it is because the MMPC relates to the statistical confidence we have in our estimated moments rather than their informativeness (see [12] for a discussion). As a result, and using such an objective function, we are able to set the detection threshold T_{det} in a completely black box manner (i.e. independent of the implementation details). Indeed, the only thing we have to guarantee is that the MMPC as computed by the objective function is significant in front of the one that would be obtained by chance, for non-informative samples. But this essentially depends on the size of the target operations. For example, the correlation between random 256-element vectors is (roughly) Gaussian-distributed[2] with mean zero. And the probability that MMPC > 0.2 by chance in this case is already below the one corresponding to three σ's (i.e. below 0.1 %). Of course, one can expect slight deviations from such an ideal behavior (e.g. so-called ghost peaks leading to non-zero mean MMPC for non-informative samples), but our next experiments will confirm that setting T_{det} to 0.2 is generally good.

Impact of W_{len}, σ_n^2 and N_i on the Detection Success. Given a detection threshold set as just explained, we can now evaluate the impact of different parameters on the success of our *find_sol* phase. In particular, the noise variance σ_n^2, number of informative pairs of samples in the traces N_i and window length W_{len} are important in this respect. As just explained, we know that given a large enough number of measurements, the MMPC criteria should become larger than 0.2 for the informative samples. But it also means that if this number of measurements is not sufficient, the moments used in MCP-DPA will not be sufficiently well estimated and the detection may fail. As usual, the main parameter influencing the estimation complexity is the noise variance σ_n^2. Yet, since we apply the objective function after projection in our PP algorithm, the size of the window W_{len} also matters here. Indeed, adding W_{len} samples with noise variance σ_n^2 implies a larger noise variance $W_{len} \times \sigma_n^2$ after projection. This is typically illustrated in the left part of Fig. 2, where we see the impact of increasing W_{len} for two noise levels ($\sigma_n^2 = 0.1$ in the top figure, $\sigma_n^2 = 2$ in the bottom one). That is, for too large noise variances or window lengths, the estimation of the

---

[2] More precise estimates can be obtained with Fisher's Z transform.

MMPC criteria is not good enough to take good decisions (i.e. is below T_{det}). In other words, more measurements are needed in this case for the PP algorithm to output meaningful results. Interestingly, we also see in the right part of the figure that adding meaningful samples in the traces (i.e. increasing N_i) quite significantly mitigates the impact of large window lengths. So intuitively, traces with multiples POIs available will better benefit from our proposed method.

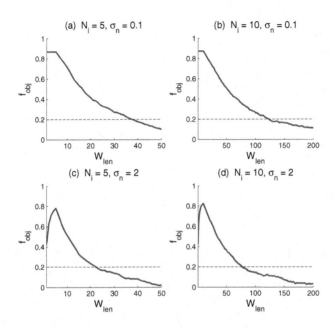

Fig. 2. Incidence of the window length W_{len} on the information detection.

Time Complexity. The previous results suggest that the complexity of PP algorithms is essentially a tradeoff between time and measurement complexities. That is, increasing the windows length should decrease their time complexity[3], but increases the noise after projection, and so the number of measurements needed to estimate the MMPC criteria with sufficient confidence. This is typically illustrated in the left part of Table 1, where we also see the benefit of having more informative samples in the traces (i.e. increasing N_i). Furthermore, the right part of the table highlights the impact of increasing the size of the traces N_s. As in a combinatorial search, the time complexity of the PP algorithm should increase quadratically with it (more generally, it depend on N_s^d with d the number of shares in the masking scheme). Yet, increasing W_{len} or N_i can make this increase quasi-linear for some (not too large) values of N_s. Besides, note that Table 1 includes all the constant factors related to the technical parameters in the

---

[3] At most linearly since the benefit of increasing the window length W_{len} saturates whenever it is not negligible in front of the number of samples in the traces N_s.

previous section, which sometimes amortizes these asymptotic predictions. Note also that this table counts the calls to the objective function for readability, but this count is not fully reflective of the PP's time complexity when changing the size of the profiling sets $\mathcal{L}_{tr}^{p}$ and $\mathcal{L}_{va}^{p}$, since larger sets also increase the complexity of each evaluation of the objective function. Yet, thanks to the parallelism of MCP-DPA attacks, the impact of these increases was limited in our experiments, leaving us with strong concrete results, as the next section will show.

Table 1. Impact of W_{len}, N_i and N_s on the average number of f_{obj} calls.

$N_s = 1000$		N_i	
		5	10
W_{len}	10	7306	4681
	20	3920	3008
	30	3266	2782
	50	-	2138
	100	-	1020
	150	-	-

	N_s		
	500	1000	2000
$W_{len} = 50, N_i = 10$	905	2138	4673

4.3 Measured Experiments

The previous simulated experiments suggest that a specialized PP algorithm can be an efficient way to find POIs in the leakage traces of masked implementations. We now would like to confirm this hope in front of a real case-study. For this purpose, we will consider the actual measurements of a first-order masked AES S-box based on table lookups [17,22]. For every pair of input/output masks (m, q), it pre-computes an S-box S^* such that $S^*(x \oplus s \oplus m) = S(x \oplus s) \oplus q$ Since this pre-computation is part of the adversary's measurements, it leads to quite memory-consuming traces of $N_s = 30,000$ samples (which would be a challenging target for a combinatorial search). Furthermore, we verified empirically that our implementation does not lead to any (easy-to-detect) first-order information leakage, by running template attacks for all the time samples, and making sure that the success rate remained negligible (which should be guaranteed by the use of independent masks m and q, in order to prevent leakages based on the transitions between the the S-box input and output). Our motivation for using this setup was twofold. First, we selected a masking countermeasure based on pre-computed tables in view of the difficulty to obtain a first-order secure implementation based on other standard masking schemes such as [21] – see [2] for a recent discussion of this problem. Second, we purposely put ourselves in a challenging scenario with large traces, without trying to compress them (e.g. by reducing the sampling frequency or through educated guess). While we agree that concrete adversaries would try to exploit these possibilities, we assume that they would not always be able to compress traces up to feasible combinatorial search, and the experiments in this section aim to reflect this possibility.

We then analyzed our set of profiling and test traces, in order to evaluate the success and efficiency of our POI detection tool. We used the same MMPC criteria and detection threshold of 0.2 as previously discussed, and selected a window length W_{len} of 25, corresponding to approximately two clock cycles in our measurements: this is the only physical intuition used in our experiments. With these parameters, it turned out that the estimation of the objective function was sufficiently accurate (for our detection threshold to make sense) with 50 profiling traces per template (i.e. 50×256 among the 500×256 measured). Based on our 1500 test traces, we then evaluated that the local search algorithm was able to return a solution within an average of 12 000 calls to f_{obj} (roughly corresponding to 7 min of execution time on our desktop computer). We then repeated this search multiple times in order to find several pairs of informative windows. We finally used these windows to launch multivariate (Gaussian) template attacks using 2, 4 and 8 dimensions. For this purpose, we selected the smallest windows (which turned out to contain 5 samples) and built templates for their mean values (so that each pair of window provided us with 2 dimensions). The results of these attacks are illustrated in Fig. 3 and confirm that our tool successfully detected POIs in this challenging case[4]. Interestingly, we see that the gain due to increased dimensionalities vanishes when moving from 4-dimension templates to 8-dimension ones. We conjecture that this mainly relates to template estimation issues. Note anyway that, as mentioned in introduction, these attacks are not aimed to be optimal from the data complexity point-of-view (since we have no guarantee to find the most informative samples). Our main goal was to provide a time-efficient POI detection tool, in a black box setting. To the best of our knowledge, previous methods for this purpose would not have been able to deal with 30,000-sample traces without an educated guess (For illustration, the product traces mentioned in footnote 1 would correspond to 900.10^6 samples).

5 Conclusions

In this work we proposed an efficient method for finding POIs in the leakage traces of cryptographic implementations. We exploit a combination of PP and local search for this purpose, and discussed the how to adapt it to the side-channel cryptanalysis problem. One of the main advantages of the method is its genericity, as it can be applied to any implementation, by simply adapting its objective function. Besides, it has very low memory requirements compared to state-of-the-art solutions and (although heuristic) works in practical time complexity. We applied our basic and specialized PP algorithms to two case studies of unprotected and 2-share masked implementations to validate our claims. Extending the specialized version to more shares would be straightforward, since this number of shares (i.e. d) is a parameter in our search algorithms.

Among the interesting open problems, we believe investigating the informativeness of the projected samples obtained with PPs in the context of protected

---

[4] For convenience, and in order to limit our measurement needs, we estimated a 4th-order success rate which corresponds to an adversary able to enumerate 2^{32} keys.

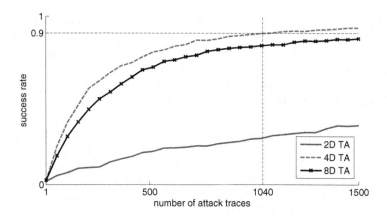

Fig. 3. 4th-order success rates of multivariate template attacks.

implementations is promising – it was essentially left out of our analysis so far. Different approaches could be considered for this purpose. One would be to refine the projection vectors, possibly based on an information theoretic objective function that would better reflect the resulting attacks' data complexity. An alternative one would be to exploit non-linear projections, e.g. inspired by the "product combining" frequently used in second-order DPA [18,26]. Yet, preliminary results suggest that non-linear projections may be hard(er) to exploit because the addition of non-informative samples when computing the objective function has higher impact on the (non-Gaussian) noise in this case. Besides, testing new objective functions that are cheap to compute and estimate, in the profiled and non-profiled settings, is another interesting research direction.

Note finally that the long version of this work includes two additional subsections with detailed discussions and comparisons with related works [6].

Acknowledgements. F.-X. Standaert is a research associate of the Belgian Fund for Scientific Research (FNRS-F.R.S.). This work has been funded in parts by the European Commission through the ERC project 280141 (CRASH).

References

1. Archambeau, C., Peeters, E., Standaert, F.-X., Quisquater, J.-J.: Template attacks in principal subspaces. In: Goubin, L., Matsui, M. (eds.) CHES 2006. LNCS, vol. 4249, pp. 1–14. Springer, Heidelberg (2006)
2. Balasch, J., Gierlichs, B., Grosso, V., Reparaz, O., Standaert, F.-X.: On the cost of lazy engineering for masked software implementations. IACR Cryptology ePrint Archive, 2014:413 (2014)
3. Batina, L., Hogenboom, J., van Woudenberg, J.G.J.: Getting more from PCA: first results of using principal component analysis for extensive power analysis. In: Dunkelman, O. (ed.) CT-RSA 2012. LNCS, vol. 7178, pp. 383–397. Springer, Heidelberg (2012)

4. Brier, E., Clavier, C., Olivier, F.: Correlation power analysis with a leakage model. In: Joye, M., Quisquater, J.-J. (eds.) CHES 2004. LNCS, vol. 3156, pp. 16–29. Springer, Heidelberg (2004)
5. Choudary, O., Kuhn, M.G.: Efficient template attacks. In: Francillon, A., Rohatgi, P. (eds.) CARDIS 2013. LNCS, vol. 8419, pp. 253–270. Springer, Heidelberg (2014)
6. Durvaux, F., Standaert, F.-X., Veyrat-Charvillon, N., Mairy, J.-B., Deville, Y.: Efficient selection of time samples for higher-order DPA with projection pursuits. IACR Cryptology ePrint Archive, 2014:412 (2014)
7. Friedman, J.H., Tukey, J.W.: A projection pursuit algorithm for exploratory data analysis. IEEE Trans. Comput. **23**(9), 881–890 (1974)
8. Gierlichs, B., Batina, L., Tuyls, P., Preneel, B.: Mutual information analysis. In: Oswald and Rohatgi [14], pp. 426–442
9. Hoos, H.H., Stützle, T.: Stochastic Local Search: Foundations and Applications. Elsevier, San Francisco (2004)
10. Mangard, S.: Hardware countermeasures against DPA – a statistical analysis of their effectiveness. In: Okamoto, T. (ed.) CT-RSA 2004. LNCS, vol. 2964, pp. 222–235. Springer, Heidelberg (2004)
11. Mangard, S., Oswald, E., Standaert, F.-X.: One for all - all for one: unifying standard differential power analysis attacks. IET Inf. Secur. **5**(2), 100–110 (2011)
12. Moradi, A., Standaert, F.-X.: Moments-correlating DPA. IACR Cryptology ePrint Archive, 2014:409 (2014)
13. Oswald, E., Mangard, S., Herbst, C., Tillich, S.: Practical second-order DPA attacks for masked smart card implementations of block ciphers. In: Pointcheval [16], pp. 192–207
14. Oswald, E., Rohatgi, P. (eds.): CHES 2008. LNCS, vol. 5154. Springer, Heidelberg (2008)
15. Poettering, B.: Fast AES implementation for Atmel's AVR microcontrollers. http://point-at-infinity.org/avraes/
16. Pointcheval, D. (ed.): CT-RSA 2006. LNCS, vol. 3860. Springer, Heidelberg (2006)
17. Prouff, E., Rivain, M.: A generic method for secure SBox implementation. In: Kim, S., Yung, M., Lee, H.-W. (eds.) WISA 2007. LNCS, vol. 4867, pp. 227–244. Springer, Heidelberg (2008)
18. Prouff, E., Rivain, M., Bevan, R.: Statistical analysis of second order differential power analysis. IEEE Trans. Comput. **58**(6), 799–811 (2009)
19. Renauld, M., Standaert, F.-X., Veyrat-Charvillon, N., Kamel, D., Flandre, D.: A formal study of power variability issues and side-channel attacks for nanoscale devices. In: Paterson, K.G. (ed.) EUROCRYPT 2011. LNCS, vol. 6632, pp. 109–128. Springer, Heidelberg (2011)
20. Reparaz, O., Gierlichs, B., Verbauwhede, I.: Selecting time samples for multivariate DPA attacks. In: Prouff, E., Schaumont, P. (eds.) CHES 2012. LNCS, vol. 7428, pp. 155–174. Springer, Heidelberg (2012)
21. Rivain, M., Prouff, E.: Provably secure higher-order masking of AES. In: Mangard, S., Standaert, F.-X. (eds.) CHES 2010. LNCS, vol. 6225, pp. 413–427. Springer, Heidelberg (2010)
22. Schramm, K., Paar, C.: Higher order masking of the AES. In: Pointcheval [16], pp. 208–225
23. Standaert, F.-X., Archambeau, C.: Using subspace-based template attacks to compare and combine power and electromagnetic information leakages. In: Oswald and Rohatgi [14], pp. 411–425

24. Standaert, F.-X., Gierlichs, B., Verbauwhede, I.: Partition *vs.* comparison side-channel distinguishers: an empirical evaluation of statistical tests for univariate side-channel attacks against two unprotected cmos devices. In: Lee, P.J., Cheon, J.H. (eds.) ICISC 2008. LNCS, vol. 5461, pp. 253–267. Springer, Heidelberg (2009)

25. Standaert, F.-X., Malkin, T.G., Yung, M.: A unified framework for the analysis of side-channel key recovery attacks. In: Joux, A. (ed.) EUROCRYPT 2009. LNCS, vol. 5479, pp. 443–461. Springer, Heidelberg (2009)

26. Standaert, F.-X., Veyrat-Charvillon, N., Oswald, E., Gierlichs, B., Medwed, M., Kasper, M., Mangard, S.: The world is not enough: another look on second-order DPA. In: Abe, M. (ed.) ASIACRYPT 2010. LNCS, vol. 6477, pp. 112–129. Springer, Heidelberg (2010)

Exploring the Resilience of Some Lightweight Ciphers Against Profiled Single Trace Attacks

Valentina Banciu$^{(\boxtimes)}$, Elisabeth Oswald, and Carolyn Whitnall

Department of Computer Science, University of Bristol,
Merchant Venturers Building, Woodland Road, Bristol BS8 1UB, UK
valentina.banciu@bristol.ac.uk

Abstract. This paper compares attack outcomes w.r.t. profiled single trace attacks of four different lightweight ciphers in order to investigate which of their properties, if any, contribute to attack success. We show that mainly the diffusion properties of both the round function and the key schedule play a role. In particular, the more (reasonably statistically independent) intermediate values are produced in a target implementation, the better attacks succeed. A crucial aspect for lightweight ciphers is hence the key schedule which is often designed to be particularly light. This design choice implies that information from all round keys can be easily combined which results in attacks that succeed with ease.

1 Introduction

In our increasingly digitally interconnected world developing secure cryptographic software is a key challenge in practice. The stakes at hand are considerable: with the advent of 'smart devices' (e.g. smart meters, smart appliances) cryptography becomes deeply embedded in consumers' everyday lives. Such devices' primary functionalities are to provide commodities to consumers. Consequently, despite the importance of cryptography, the resources remaining to implement it may be very limited. This requirement for lightweight cryptography has inspired many new designs, such as PRESENT [3] (recently standardised as ISO/IEC 29192-2:2012), KLEIN [8] and LED [9] among many others. Some were designed with suitability for hardware implementations in mind (such as PRESENT), but by and large all of them are also considered for implementations in software on small microprocessors as found in typical smart devices.

Deeply embedded systems are by nature very exposed to side-channel adversaries: their market is such that millions of these devices can be found 'in the wild' as part of applications, but many of the processors that are used in smart devices can be picked up off the shelf for only a small cost. This makes them ideal targets for the most sophisticated side-channel attacks, which are based on profiling. Profiling can be used both in simple and differential style attacks. Differential style attacks exploit leaking information associated with different plaintext data but a fixed key and it is well understood that the choice of e.g. the substitution function has an important impact on the vulnerability of a cipher [14].

© Springer International Publishing Switzerland 2015
S. Mangard and A.Y. Poschmann (Eds.): COSADE 2015, LNCS 9064, pp. 51–63, 2015.
DOI: 10.1007/978-3-319-21476-4_4

In this paper we aim to investigate the security of specific implementations of lightweight ciphers against profiled single trace attacks. In particular, we are interested to discover which properties, if any, contribute to the general vulnerability of a lightweight cipher against this particular type of adversary.

Our Contribution. Using a pragmatic approach to assess the resilience of ciphers against profiled single trace attacks, we conduct several experiments for four ciphers: AES (to provide a well-known baseline for comparison), PRESENT (to see how ciphers which are attractive for hardware implementations fare when implemented in software), KLEIN and LED (sharing some features with AES and PRESENT while aiming at resource minimisation). All experiments assume the same underlying architecture (8-bit) and are based on publicly available implementations of the investigated ciphers [7]. Apart from providing a common denominator, we regard this suite as relevant to a high proportion of real-life cases since algorithm implementations are done in a 'natural' way. Note that we work with simulated traces, which however does not imply perfect measurements (see Sect. 3 for details on how inherent noise is taken into account).

Our method, which we describe in Sect. 4.1, can be regarded as a basic tool for evaluating the resilience of specific cipher implementations against single trace attacks. As such, we draw inspiration from [1,10] which independently suggested using a pragmatic key enumeration approach. However, instead of enumerating single subkeys we test multiple candidates at the same time, thus achieving 100 % success rate and running time of under 5 min for all test cases, which represent significant improvements.

We can show that the impact of choice of substitution boxes is negligible w.r.t. profiled single trace attacks and equally the impact of the global diffusion characteristics over a single round have shown no impact in the case of the four studied ciphers. We find however that simply the number of statistically independent intermediate steps (i.e. the 'attack surface') that are required on the architecture is a good predictor for the vulnerability of a cipher against profiled single trace attacks. This shows that ciphers that can be elegantly described on a variety of architectures will be most resilient to such attacks as they provide a smaller attack surface.

Related Work. As mentioned above, this work is strongly linked to [1,10], but introduces significant improvements in terms of running time and success rate. As such, and because the output of our attack is a reduced key space, our method represents an uncostly means of evaluating the worst-case scenario of single-trace attacks such as solver-aided algebraic side-channel attacks [4,13,16]. The cited papers focus on AES and PRESENT encryption, under the Hamming weight leakage model. Further leakage models are considered in [12,15]. We are not aware of single trace attacks on KLEIN and LED. The security of the (unprotected) AES key schedule algorithm has been studied in [11,17]. More recently, [5] described attacks on masked implementations in an ideal scenario where noise is practically negligible. We are not aware of single trace attacks on the KLEIN or PRESENT key schedule.

2 Overview of Ciphers, Notation and Implementation Characteristics

The ciphers in our suite are instances of substitution-permutation networks (SPN), and therefore have similar components in their encryption functions. In particular, three common types of operations are utilised by all four ciphers:

- the key addition (bitwise xor between the round key and current state), denoted by AddRoundKey;
- the n-bit substitution (a highly non-linear transformation which acts upon groups of n bits substituting them via a lookup table called the S-box, with $n \in \{4, 8\}$ fixed), represented by SubBytes, SubNibbles, sBoxLayer and SubCells;
- finally, the byte mixing (a linear transformation acting on sets of bits), namely MixColumns, MixNibbles, pLayer and MixColumnsSerial.

Except PRESENT, the chosen ciphers utilise explicit byte (or 4-bit) renumbering functions, i.e. ShiftRows and ShiftNibbles. Additionally, the AddRoundConstant of LED xors the state with a round constant (can be done at the same time as the key addition).

AES. The Advanced Encryption Standard (AES) is a symmetric block cipher with a fixed block size of 128 bits and a variable key size of 128, 192, or 256 bits respectively corresponding to 10, 12 and 14 encryption rounds. Throughout this document we refer to the 128-bit key variant simply as AES. At the beginning of the encryption process, the plaintext is xor-ed with the secret key. Subsequently, each encryption round bar the last one consists of the successive application of SubBytes, ShiftRows, MixColumns and AddRoundKey; the last encryption round skips MixColumns. The key expansion is elegant. It reuses components from the round function and operates on so-called 'words'.

KLEIN. KLEIN is an AES-like lightweight block cipher, supporting a fixed 64-bit state and 64, 80, or 96-bit keys for 12, 16, respectively 20 rounds. Throughout this document we refer to the 64-bit key variant simply as KLEIN. Each encryption round consists of the successive application of AddRoundKey, SubNibbles, RotNibbles and MixNibbles. A final key addition is performed after the encryption rounds. Although the order of operations inside an encryption round is different, their succession starting from the beginning of the encryption process is the same as with AES; moreover, the MixNibbles of KLEIN is identical to the MixColumns of AES. The key expansion process is fairly simple, each new key byte depending on exactly two bytes from a previous key.

PRESENT. PRESENT consists of 31 rounds, has a 64-bit block size and 80 or 128-bit keys. Throughout this document we refer to the 80-bit key variant simply as PRESENT. An encryption round consists of the key addition AddRoundKey, followed by the substitution and permutation layers, sBoxLayer and pLayer. The permutation layer is designed to match the effects of the combination between ShiftRows and MixColumns of AES. A final key addition is performed after the encryption rounds. The key schedule is fairly minimal: each new round key is derived from the previous key via a bit rotation, a single application of the S-box and a single xor with a round constant.

LED. LED accepts a 64-bit block and 64-bit or 128-bit keys, and consists of 32, respectively 48 rounds. Throughout this document we refer to the 64-bit key variant simply as LED. The structure of an encryption round is the succession of AddRoundConstant, SubCells, ShiftRows and MixColumnsSerial. Then, four rounds make a step, and the encryption process consists of adding the round key and performing a step for a total of 8 times, followed by a final key addition. LED has no key expansion algorithm, and the secret key is used as a round key in each round.

As can be observed, the structure of the encryption algorithms is highly similar: first, because homologous subroutines are used, and second, because the order of the subroutines is virtually the same. We note that the first AddRoundConstant of LED can be performed before the key addition, i.e. directly on the plaintext.

2.1 Implementation Characteristics

SPA attacks are usually studied in the context of software implementations on serial microprocessors. Typical power models that are found in practice are the Hamming weight (HW) and the Hamming distance (HD). Leakages of this kind are observed mainly because of intermediate values being written to or read from memory.

For AES, KLEIN and PRESENT, we target publicly available 8-bit implementations, available at http://perso.uclouvain.be/fstandae/lightweight_ciphers/. In the case of LED, the publicly available implementations can be found at http://led.crypto.sg/software. In the remainder of this section, we further discuss implementation details of the substitution and byte mixing layers; we consider that implementing the key or round constant addition is fairly straightforward and does not require clarification.

The Byte Substitution Layer. As described in Sect. 2, KLEIN, PRESENT and LED use 4-bit lookup tables. In order to optimally fit on 8-bit architectures two consecutive nibbles (2×4 bits) are considered a unit and new lookup tables are built [7].

The Byte Mixing Layer. For all ciphers this is the most demanding component w.r.t. efficient implementations on an 8-bit platform. We briefly explain the approaches taken by the different ciphers in turn.

MixColumns (used in AES and KLEIN). The implementation that we are targeting [7] follows the specification of the original AES proposal [6], and is given in Algorithm 1, where the index i wraps around $1 \ldots 4$, i.e. $i + 1 = 1$ if $i = 4$. As mentioned before, KLEIN uses the same component but calls it `MixNibbles`.

`pLayer` (used in PRESENT). A naïve implementation of `pLayer` would consist of storing a table that describes the bit-level permutation. However, this takes up a considerable amount of memory and is unnecessary because the `pLayer` permutation is highly structured. The targeted implementation [7] uses this property and does not require any table-lookup. Algorithm 2 shows how the implementation that we attack applies the permutation to half of the state.

`MixColumnsSerial` (used in LED). In the specification paper of LED [9] the authors suggest an 8-bit implementation of `MixColumnsSerial` using lookup tables, see Algorithm 4. Each element is regarded as part of $GF(2^4)$ with the underlying polynomial for field multiplication given by $x^4 + x + 1$.

3 Assessing the Vulnerability to Profiled Single Trace Attacks

Profiling can be used in both differential and simple side-channel analysis, and it is well understood that in general it improves attack efficiency. However, for some methods it is considered essential: so-called single trace attacks such as e.g. SPA against the AES key schedule [11] or algebraic side-channel attacks (ASCA, [16]) require the extraction of leakage values from traces and their assignment to specific intermediate values in a cipher's implementation. This is a demanding requirement which by and large can *only* be satisfied by profiling an implementation, i.e. by extracting leakage models for all exploitable intermediate values, and then using them during attacks. Such profiles hence contain information about 'when' an intermediate leaks (i.e. they have information about the timing of instructions) as well as 'how' (i.e. the leakage model itself).

Whilst it remains an open problem to deal with errors related to when leakages occur in profiling attacks, there has been some progress w.r.t. dealing with errors that result from matching the profiling information to new traces. For the two approaches to single trace attacks (which we call pragmatic SPA and ASCA) previous work introduced the notion of a 'set size' [1,12,13] to capture the impact of noise on attacks using profiling information. The larger the set size, the less certain we are about the assignment of a leakage value to an intermediate, e.g. a set size of three implies that for a certain intermediate we have three possible leakage values as a result of using the profiling information. To assess then the vulnerability to profiled single trace attacks we are hence interested to experiment with different set sizes. For a study on the practical requirements of extracting side-channel information to this end, we refer the reader to [2].

As previously mentioned there are broadly two types of single trace attacks in the literature at present. Pragmatic SPA-style attacks were described early on [11] and essentially consist of enumerating key candidates by exploiting the leakage information across a single trace. This enumeration is by and large manually implemented and the result of such attacks is hence information about the

Algorithm 1. MixColumns 8-bit implementation algorithm

Input: in_1, in_2, in_3, in_4
Output: $out_1, out_2, out_3, out_4$
1: $Tmp \leftarrow in_1 \oplus in_2 \oplus in_3 \oplus in_4;$
2: **for** $i = 1 \rightarrow 4$ **do**
3:     $Tm \leftarrow in_i \oplus in_{i+1};$
4:     $Tm \leftarrow \texttt{xtime}(Tm);$
5:     $out_i \leftarrow in_i \oplus Tm \oplus Tmp;$
6: **end for**

Algorithm 2. pLayer (permuting 32 bits)

Input: in_1, in_2, in_3, in_4
Output: $out_1, out_2, out_3, out_4$
1: $carry_1 \leftarrow 0; carry_2 \leftarrow 0;$
2: $out_1 \leftarrow 0; out_2 \leftarrow 0; out_3 \leftarrow 0; out_4 \leftarrow 0;$
3: **for** $i = 4 \rightarrow 1$ **do**
4:     $iTmp \leftarrow in_i;$
5:     pLayerByte($iTmp, out_1, out_2, out_3, out_4, carry_2$);
6:     $carry_2 \leftarrow carry_1;$
7: **end for**

Algorithm 3. pLayerByte($iTmp, out_1, out_2, out_3, out_4, carry_2$)

Input: $iTmp, out_1, out_2, out_3, out_4, carry_2$
Output: $out_1, out_2, out_3, out_4, carry_1$
1: $carry_1 = \text{mod}(iTmp, 2);$
2: $iTmp = \text{floor}(iTmp/2) + carry_2 \times 128;$
3: **for** $repeat = 1 \rightarrow 2$ **do**
4:     **for** $i = 1 \rightarrow 4$ **do**
5:         $carry_2 = \text{mod}(out_i, 2);$
6:         $out_i = \text{floor}(out_i/2) + carry_1 * 128;$
7:         $carry_1 = \text{mod}(iTmp, 2);$
8:         $iTmp = \text{floor}(iTmp/2) + carry_2 * 128;$
9:     **end for**
10: **end for**

Algorithm 4. MixColumnsSerial 8-bit implementation algorithm

Input: in_1, in_2, in_3, in_4
Output: $out_1, out_2, out_3, out_4$
1: $out_1 \leftarrow 4 \times in_1 \oplus 1 \times in_2 \oplus 2 \times in_3 \oplus 2 \times in_4;$
2: $out_2 \leftarrow 8 \times in_1 \oplus 6 \times in_2 \oplus 5 \times in_3 \oplus 6 \times in_4;$
3: $out_3 \leftarrow \text{B} \times in_1 \oplus \text{E} \times in_2 \oplus \text{A} \times in_3 \oplus 9 \times in_4;$
4: $out_4 \leftarrow 2 \times in_1 \oplus 2 \times in_2 \oplus \text{F} \times in_3 \oplus \text{B} \times in_4;$

size of the key space left to search through to find the secret key. In contrast, ASCA was developed later [16] and essentially feeds side-channel information in addition to plain and ciphertext to a solver which will then return the secret key unless it halts. ASCA was hoped to be more efficient as solvers are sophisticated software tools. It should be evident however that to assess the vulnerability of ciphers they are less suitable than pragmatic attacks: they either return the key or produce no information. In contrast pragmatic attacks allow to assess the size of the remaining key space (i.e. remaining after all side-channel information has been used to prune the overall key space) and so give us some information about how much the side-channel information has helped. This becomes particularly useful when considering larger set sizes: recent work [1] shows how pragmatic attacks can produce useful information for set sizes up to 5, whereas ASCA is unable to cope with such large set sizes. Consequently it seems most appropriate to use pragmatic attacks as an evaluation tool with increasing set sizes (we report results for set sizes up to 5).

In the following sections we investigate three important characteristics in turn. Firstly, is there any high level difference between the round functions of ciphers w.r.t. profiled single trace attacks? This means we look at attacks that only use some selected intermediates corresponding to the key components of any substitution-permutation network. Next, we investigate how the inclusion of additional intermediate leakages changes attack outcomes. Thirdly, we study how key schedule characteristics impact on the vulnerability.

For these attacks, we generated a set of 100 random 16-byte plaintext and ciphertext pairs, which are the fixed inputs for the cipher suite; when a smaller block is required, the pairs are truncated (i.e., for e.g. KLEIN the secret keys will consist of the first 8 bytes of each 16-byte key). Note that our attack actually utilises a single trace, thus the reported results are in fact averaged over 100 experiments.

4   Attacking Selected Intermediates from a Single Encryption Round

There are four steps across all ciphers in which side-channel relevant computations occur:

- Loading the secret key from memory (we assume the plaintext is always known);
- Performing the key addition (and the xor-ing with the round constant if the case);
- Performing the substitution;
- Computing the output of the byte mixing layer (i.e., MixColumns, MixNibbles, pLayer, MixColumnsSerial).

In the remainder of this section we describe our attack methodology and show how the side-channel information reduces the key space when considering these four steps for a single encryption round. Because all ciphers effectively act on the state in some block-wise manner we first explain what our basic 'block' is. In AES all but `MixColumns` act on individual bytes of the state. `MixColumns` operates on columns which implies that a suitable block would be a column (i.e. 4 bytes). Studying how a pragmatic SPA reduces the key space with regards to this block allows us to conclude on the result of the entire key because the blocks are independent. It is easy to see that such a 4-byte block is also an appropriate unit for the other ciphers when implemented on an 8-bit platform. Consequently we settled for this choice and the tables in this section show the reduction of the subkey space for a block (i.e. from 2^{32}). Note that this definition overrides the one in Sect. 2 without contradicting or hindering any of the inherent cipher properties.

4.1 Attack Strategy

As mentioned in Sect. 1, our attack is derived from [1] and therefore similarly consists of two phases: first, extracting four independent sets of key (byte) values based on the side-channel information up to the byte mixing layer, and second, linking the extracted key bytes into 4-byte keys based on the information from the byte mixing layer. Indeed, we also bui ld and use 8-bit tables that enable us to directly extract possible key values based on the known plaintext and side-channel information from the S-box for all ciphers. However, for the second part, instead of then enumerating the (4-byte) keys and testing each one sequentially as in [1], we generate all possible keys (corresponding to a block) as the Cartesian product of the previously derived sets and simulate their action on the inputs. With this we are able to reject several key candidates at the same time based on the side-channel information, which allows us to report a short running time for our attacks (under 5 min, but under one second for set sizes up to 2) and a success rate of 100 % (previous attacks were liable to run out of memory, or to fail to complete within a fixed time interval, e.g. 48 h). This is a significant improvement to previous work.

All our experiments ran on a regular PC equipped with a Intel Core i7-2600s processor at 2.80 GHz and 4 GB of RAM.

4.2 Exploiting the 'Basic' Attack Surface

We first consider that the sole available side-channel information is related to the input and output values of the four steps outlined at the beginning of this section. Then, Table 2a summarizes the reduced subkey space for a block of the ciphers. It appears that the size of the reduced subkey space strongly depends on the set size, and less so on the specific cipher particularities (i.e. the quality of the S-box or byte mixing function are by and large irrelevant). Of course the overall key space of the ciphers is different and hence there is an additional penalty for AES as it requires to replicate the attack for more subkeys than the other ciphers.

4.3   Exploring the Impact of Increased Numbers of Intermediates

A natural question to ask is whether more leaking intermediate values will make an implementation more vulnerable, and if so, whether there is any clear relationship between the number of leaking intermediates and the increase in vulnerability. We hence took all of the intermediate values that occur in our implementations into account (i.e., the 'maximum' attack surface). Previous work (e.g., [15]) studied the impact of using more intermediate values by targeting more encryption rounds. Note that we are still focusing on a single round. We now explain for each of the ciphers in turn what and how many intermediates our implementations offer.

The implementation of MixColumns given in Algorithm 1 leads to a set of 17 intermediate values, as follows: 4 corresponding to computing $in_i \oplus in_{i+1}$, 4 corresponding to computing $\texttt{xtime}(Tm)$, 8 corresponding to computing $iv = in_i \oplus Tm$ and $iv \oplus Tmp$ (where iv is an auxiliary intermediate value), and finally one corresponding to computing Tmp (n.b.: $in_i \oplus in_{i+1}$ have already been computed, therefore a single new value is leaked when computing Tmp).

The implementation of pLayer given in Algorithm 2 leaks as follows: the pLayerByte procedure leaks $2 + 2 \times 4 \times 4$ byte values, and is repeated a total of 4 times, therefore leading to 116 intermediate values. Note, however, that this set consists of values that differ in a single bit. This implies that although many intermediate values are produced, they are highly correlated. Consequently, given the 8-bit architecture that we work on, we effectively observe multiple copies of only 8 intermediates.

The implementation of MixColumns given in Algorithm 4 leaks 7 intermediate values (4 for the table lookup, and 3 for computing the binary xor operations) for each output byte, thus leaking a total of 28 leakage points.

Table 1 gives an overview of the number of leaking intermediates per cipher. We listed only 8 intermediates for PRESENT because of the evident high correlation between the intermediates. We note that also for MixColumns there will be some correlated intermediates due to the fact that e.g. the final sum is computed by xor-ing. From this table, we would expect to see that LED should suffer most from including these additional intermediates, followed by AES and KLEIN.

Table 2b shows the results when incorporating the additional intermediates into the attack. All ciphers show that the simple intuition that more statistically independent intermediates provide more efficient attacks is true. The results for PRESENT also provide a clear example for the importance of having statistically independent intermediate values: although the total number of used intermediates is almost 7 times as large as with AES and KLEIN, the sizes of the respective reduced key spaces remain comparable.

It is thus evident that the relationship between the number of intermediates and the attack efficiency does not follow a simple linear rule. This is most likely because different intermediates are not entirely independent from each other and so they do not equally contribute additional information.

Table 1. Size of the attack surface (i.e., number of leaked intermediate values) corresponding to the diffusion layer

Attack surface \ Cipher	AES	KLEIN	PRESENT	LED
'Basic'	4	4	4	4
'Maximum'	21	21	12	32

Table 2. Reduced key space when targeting the encryption function

(a) Targeting the 'basic' attack surface

Cipher \ Setsize	HW model					HD model				
	$s=1$	$s=2$	$s=3$	$s=4$	$s=5$	$s=1$	$s=2$	$s=3$	$s=4$	$s=5$
AES	3	2^{10}	2^{20}	2^{23}	2^{25}	30	2^{15}	2^{22}	2^{25}	2^{26}
KLEIN	3	2^{9}	2^{12}	2^{18}	2^{23}	90	2^{15}	2^{22}	2^{24}	2^{26}
PRESENT	23	2^{11}	2^{19}	2^{23}	2^{25}	60	2^{15}	2^{22}	2^{24}	2^{25}
LED	2	2^{10}	2^{18}	2^{21}	2^{24}	35	2^{16}	2^{21}	2^{23}	2^{25}

(b) Targeting the 'maximum' attack surface

Cipher \ Setsize	HW model					HD model				
	$s=1$	$s=2$	$s=3$	$s=4$	$s=5$	$s=1$	$s=2$	$s=3$	$s=4$	$s=5$
AES	1	1	2^{10}	2^{18}	2^{24}	1	1	2^{12}	2^{19}	2^{24}
KLEIN	1	1	2^{7}	2^{12}	2^{20}	1	1	2^{9}	2^{14}	2^{21}
PRESENT	1	1	2^{8}	2^{12}	2^{20}	1	1	2^{10}	2^{13}	2^{22}
LED	1	1	2^{5}	2^{11}	2^{19}	1	1	2^{7}	2^{13}	2^{20}

5   Attacking the Key Expansion

The key expansion algorithms are substantially different w.r.t. their diffusion properties. We hence briefly run through them in turn to explain what attack strategies are possible. Let the shorthand RK stand for round key. We use $RK_i(j)$ for the j-th byte of the i-th round key.

Our principal contribution in this section is describing single trace attacks on the key schedule of KLEIN and PRESENT. We remind the reader that the first attack on the key schedule of AES (which we reproduce here as well, considering larger set sizes) has been described in [11]. As mentioned in Sect. 2, LED uses the same secret key for all encryption rounds.

5.1 Attack Strategies

AES. The particularities of AES make it possible to target 5-byte subkeys and a set of 5 consecutive round keys, as first described in [11]. Thus, the results that we report are on one round key (as in [1]) and on 5 rounds (as in [11]) considering sets of up to 5 values. Because of the properties of the AES key expansion, attacks utilising all 10 round keys become computationally demanding for larger set sizes [17].

KLEIN. The KLEIN key schedule is relatively simple to attack. One can target 2-byte subkeys and use as many round keys as available (see Fig. 1). Thus, we list results for the attack utilising one round key, all round keys and half of the round keys.

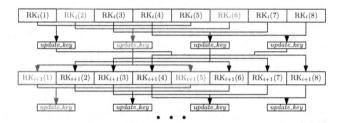

Fig. 1. Targeted KLEIN subkey

PRESENT. The key expansion of PRESENT is almost non-existent: each round key is derived from the previous via a cyclic shift of bits, a single application of the 4-bit S-box and an `xor`-ing with a round constant. Thus, reporting an attack on a single round key makes no sense, and we give the results on the full key schedule.

5.2 Attack Outcomes

Table 3 contains the outcomes of all attacks following the previously outlined attack strategies. We can observe that the diffusion properties, which impact on how much information from the key schedule we can incorporate given our computational abilities, play a significant role in attack outcomes. Consider AES for example: for larger set sizes we can only utilise the leakages from the first five round keys. Consequently the remaining key space is considerable, albeit much reduced. KLEIN in contrast is very vulnerable as we can effectively utilise all leakages across the key schedule and so we can tolerate high set sizes. PRESENT not only has a weak diffusion but also highly correlated intermediates in its key schedule and hence suffers much less from the lack of diffusion: it remains more resilient to attacks utilising leaks from the key expansion.

Table 3. Reduced key space when targeting the key expansion

(a) AES

Set size / # RK	HW model					HD model				
	$s=1$	$s=2$	$s=3$	$s=4$	$s=5$	$s=1$	$s=2$	$s=3$	$s=4$	$s=5$
1[1]	2^{58}	2^{74}	2^{95}	2^{106}	2^{115}	2^{60}	2^{75}	2^{99}	2^{107}	2^{118}
5	10	2^{15}	2^{35}	2^{58}	n.a.	30	2^{17}	2^{37}	2^{55}	n.a.
11[17]	1	n.a.	n.a.	n.a.	n.a.	n.a.	n.a.	n.a.	n.a.	n.a.

(b) KLEIN

Set size / # RK	HW model					HD model				
	$s=1$	$s=2$	$s=3$	$s=4$	$s=5$	$s=1$	$s=2$	$s=3$	$s=4$	$s=5$
1	2^{35}	2^{45}	2^{50}	2^{57}	2^{60}	2^{40}	2^{48}	2^{55}	2^{57}	2^{61}
6	2^{8}	2^{15}	2^{35}	2^{45}	2^{55}	2^{12}	2^{21}	2^{37}	2^{49}	2^{57}
12	1	2^{4}	2^{20}	2^{32}	2^{45}	1	2^{4}	2^{22}	2^{37}	2^{50}

(c) PRESENT

Set size / # RK	HW model					HD model				
	$s=1$	$s=2$	$s=3$	$s=4$	$s=5$	$s=1$	$s=2$	$s=3$	$s=4$	$s=5$
31	2^{10}	2^{16}	2^{45}	2^{60}	2^{73}	2^{14}	2^{16}	2^{45}	2^{60}	2^{73}

6 Conclusion

In this paper we investigated, using pragmatic SPA attacks as an evaluation tool, how different lightweight ciphers compare with regards to their vulnerability against profiled single trace attacks. The aim was to tease out which of their properties, if any, have an influence on the efficiency of such attacks.

We found that for both the encryption round function and the key schedule the diffusion properties were decisive for attack success: the more reasonably statistically independent intermediate values occur in a concrete implementation, the better a profiled single trace attack could fare. This means that such attacks not only reduce the key space further for a subsequent brute force search, but also cope better with erroneous side-channel information i.e. they can tolerate larger set sizes. The fact that most lightweight ciphers feature a particularly lightweight key schedule with little diffusion means that attacks can easily exploit the information from all round keys; this implies stronger attacks.

Acknowledgements. V. Banciu has been supported by EPSRC via grant EP/H049606/1. E. Oswald and C. Whitnall have been supported in part by EPSRC via grant EP/I005226/1. The authors would like to thank the anonymous reviewers for the useful comments and suggestions.

References

1. Banciu, V., Oswald, E.: Pragmatism vs. elegance: comparing two approaches to simple power attacks on AES. In: Prouff, E. (ed.) COSADE 2014. LNCS, vol. 8622, pp. 29–40. Springer, Heidelberg (2014)
2. Banciu, V., Oswald, E., Whitnall, C.: Reliable information extraction for single trace attacks. IACR ePrint Arch. **2015**, 45 (2015)
3. Bogdanov, A.A., Knudsen, L.R., Leander, G., Paar, C., Poschmann, A., Robshaw, M., Seurin, Y., Vikkelsoe, C.: PRESENT: an ultra-lightweight block cipher. In: Paillier, P., Verbauwhede, I. (eds.) CHES 2007. LNCS, vol. 4727, pp. 450–466. Springer, Heidelberg (2007)
4. Carlet, C., Faugère, J.-C., Goyet, C., Renauld, G.: Analysis of the algebraic side channel attack. JCEN **2**(1), 45–62 (2012)
5. Clavier, C., Marion, D., Wurcker, A.: Simple power analysis on AES key expansion revisited. In: Batina, L., Robshaw, M. (eds.) CHES 2014. LNCS, vol. 8731, pp. 279–297. Springer, Heidelberg (2014)
6. Daemen, J., Rijmen, V.: Rijndael for AES. In: AES Candidate Conference, pp. 343–348, (2000)
7. Eisenbarth, T., Kumar, S., Paar, C., Poschmann, A., Uhsadel, L.: A survey of lightweight-cryptography implementations. IEEE Des. Test Comput. **24**(6), 522–533 (2007)
8. Gong, Z., Nikova, S., Law, Y.W.: KLEIN: a new family of lightweight block ciphers. In: Juels, A., Paar, C. (eds.) RFIDSec 2011. LNCS, vol. 7055, pp. 1–18. Springer, Heidelberg (2012)
9. Guo, J., Peyrin, T., Poschmann, A., Robshaw, M.: The LED block cipher. In: Preneel, B., Takagi, T. (eds.) CHES 2011. LNCS, vol. 6917, pp. 326–341. Springer, Heidelberg (2011)
10. Guo, S., Zhao, X., Zhang, F., Wang, T., Shi, Z.-J., Standaert, F.-X., Ma, C.: Exploiting the incomplete diffusion feature: a specialized analytical side-channel attack against the AES and its application to Microcontroller implementations. IEEE Trans. Inf. Forensics and Secur. **9**(6), 999–1014 (2014)
11. Mangard, S.: A simple power-analysis (SPA) attack on implementations of the AES key expansion. In: Lee, P.J., Lim, C.H. (eds.) ICISC 2002. LNCS, pp. 343–358. Springer, Heidelberg (2003)
12. Oren, Y., Renauld, M., Standaert, F.-X., Wool, A.: Algebraic side-channel attacks beyond the Hamming weight leakage model. In: Prouff, E., Schaumont, P. (eds.) CHES 2012. LNCS, vol. 7428, pp. 140–154. Springer, Heidelberg (2012)
13. Oren, Y., Wool, A.: Tolerant algebraic side-channel analysis of AES. IACR ePrint Arch. **2012**, 92 (2012)
14. Prouff, E.: DPA attacks and S-boxes. In: Gilbert, H., Handschuh, H. (eds.) FSE 2005. LNCS, vol. 3557, pp. 424–441. Springer, Heidelberg (2005)
15. Renauld, M., Standaert, F-X.: Representation-, leakage- and cipher- dependencies in algebraic side-channel attacks. In: Industrial Track of ACNS, pp. 1–18 (2010)
16. Renauld, M., Standaert, F.-X., Veyrat-Charvillon, N.: Algebraic side-channel attacks on the AES: why time also matters in DPA. In: Clavier, C., Gaj, K. (eds.) CHES 2009. LNCS, vol. 5747, pp. 97–111. Springer, Heidelberg (2009)
17. VanLaven, J., Brehob, M., Compton, K.J.: A computationally feasible SPA attack on AES via optimized search. In: Sasaki, R., Qing, S., Okamoto, E., Yoshiura, H. (eds.) Security and Privacy in the Age of Ubiquitous Computing. IFIP Advances in Information and Communication Technology, pp. 577–588. Springer, New York (2005)

Two Operands of Multipliers
in Side-Channel Attack

Takeshi Sugawara$^{(\boxtimes)}$, Daisuke Suzuki, and Minoru Saeki

Mitsubishi Electric Corporation, Kamakura, Japan
sugawara.takeshi@bp.mitsubishielectric.co.jp

Abstract. The single-shot collision attack on RSA proposed by Hanley et al. is studied focusing on the difference between two operands of multipliers. There are two consequences. Firstly, designing order of operands can be a cost-effective countermeasure.We show a concrete example in which operand order determines success and failure of the attack. Secondly, countermeasures can be ineffective if the asymmetric leakage is considered. In addition to the main results, the attack by Hanley et al. is extended using the signal-processing technique of the big mac attack. An experimental result to successfully analyze an FPGA implementation of RSA with the multiply-always method is also presented.

Keywords: RSA · Side-channel attack · Collision attack · Montgomery multiplication

1 Introduction

Side-channel attacks use unintentional information leakage from secure chips to compromise their security. New attacks and countermeasures have been studied for years since the first attack was discovered in 90s [1].

Side-channel attacks are divided into multiple- and single-shot attacks depending on the number of traces used. The first side-channel attack of RSA presented by Kocher et al. is a single-shot attack [1]. A typical modular exponentiation algorithm makes conditional branch between multiplication and squaring depending on a bit of the secret exponent. Kocher et al. showed that the multiplication and squaring are distinguishable by analyzing a power trace, and thus the secret exponent can be revealed.

Conditional branch is easily exploitable and thus should be removed. The multiply-always method in Algorithm 1 is a well-known method to implement RSA without data-dependent branch. Even after data-dependent branch is removed, there is residual side-channel leakage correlated to the operands of the multiplication and squaring [4]. The residual leakage is weak [5], but the emerging new attacks can exploit it. Recently, successful single-shot attacks on FPGA implementations were reported [6–8]. The successful single-shot attacks have a large impact in designing secure implementations. That is because suppressing the residual leakage requires a lot of effort. Hanley et al. suggest that

© Springer International Publishing Switzerland 2015
S. Mangard and A.Y. Poschmann (Eds.): COSADE 2015, LNCS 9064, pp. 64–78, 2015.
DOI: 10.1007/978-3-319-21476-4_5

a multiplier-level countermeasure [9–11] is needed for the suppression. In this paper, leakage from multipliers is studied. In contrast to the previous works on leakage from multipliers [9,12,13], asymmetry between operands of multipliers is focused. The contributions of this paper are summarized as follows.

A1 The single-shot collision attack on RSA proposed by Hanley et al. is studied focusing on the difference between two operands of multipliers. The asymmetry is reasoned by the Booth recoding and operand scanning.
A2 It is shown that designing order of operands can be a cost-effective countermeasure.
A3 It is shown that some countermeasure become ineffective when the asymmetric leakage is considered.

In addition to the above main results, there are two additional contributions.

B1 The single-shot attack by Hanley et al. [8] is extended using the technique of the big mac attack [12].
B2 An experimental result to successfully analyze an FPGA implementation of RSA with the multiply-always method is presented.

The paper is organized as follows. The conventional internal collision attacks are reviewed in Sect. 2, followed by the proposed extension of the attack by Hanley et al. Difference of operands of multipliers is discussed in Sect. 3. The experimental results are shown in Sect. 4. In the section, the attack in Sect. 2 is applied to an FPGA implementation with various operand orders. The experimental result are discussed in Sect. 5. Section 6 is a concluding remark.

2   Single-Shot Collision Attack

Firstly, conventional attacks are briefly reviewed. Then, the two most relevant attacks namely (i) the multiple-shot attack by Witteman et al. [14] and (ii) the single-shot attack by Hanley et al. [8] are described in detail. Finally, the proposed extension of the attack by Hanley et al. is described.

2.1   Conventional Single-Shot Attacks

Simple Power Analysis (SPA) [1]. As described in the introduction, Kocher et al. proposed the first single-shot attack on RSA. The binary method used for modular exponentiation is targeted. In the binary method, there is a branch between square and multiplication depending a bit of the secret exponent. Kocher et al. showed that the path taken in the branch can be distinguished by analyzing a power trace.

Big Mac Attack (BMA) [12]. Walter proposed BMA to attack another modular exponentiation algorithm called the window method [12]. The idea is to compare two segments of a power trace in order to find collision. In addition,

---

Algorithm 1. Multiply-Always Method with Left-To-Right Scanning

---

Input: Message M, Modulo N, Secret exponent $d = (d_{t-1}, \cdots, d_0)_2$
Output: Ciphertext M^d
  1: $R_0 \leftarrow 1$
  2: **for** $j = t - 1$ downto 0 **do**
  3:     $\overline{d_j} \leftarrow 1 - d_j$
  4:     $R_0 \leftarrow R_0^2 \mod N$
  5:     $R_{\overline{d_j}} \leftarrow R_0 \times M \mod N$
  6: **end for**
  7: Return R_0

---

a sophisticated signal-processing technique is introduced to improve the performance of the comparison. Firstly, the segment is split into multiple sub-traces. Then the sub-traces are averaged together to make a processed segment. Signal-to-noise ratio (SNR) is improved by the processing. Finally the processed segments are compared. The feasibility of the attack is proved with simulation [12]. However, no practical result has been reported as described in [15].

Horizontal Correlation Power Analysis (HCPA) [9]. HCPA proposed by Clavier et al. is a successor of BMA [9]. In the attack, a single trace is split into many sub-traces in the same manner as BMA. Then, a multiple-shot attack is mounted to the virtual multiple traces. There are experimental results successfully attacking software implementations [9].

Clustering-Based Attacks [6]. Heyszl et al. proposed an attack using the k-means clustering [6]. That is then improved by Perin et al. [7]. In those attacks, segments of traces are classified into two groups using the k-means algorithm. The two groups expectedly correspond to 0 and 1 of secret bits. FPGA implementations are defeated by the attacks [6,7]. Notably, Perin et al. successfully attacked an FPGA implementation with a multiplier-level countermeasure (the leak resilient arithmetic [11]) by exploiting the remaining first-order leakage.

2.2 Multiple-Shot Internal Collision Attack by Witteman et al. [14]

Witteman et al. proposed a new multiple-shot attack which exploits collision between consecutive operations i.e., internal collision [14]. The attack on the multiply-always method (Algorithm 1) is described.

    The consecutive multiplication and squaring in Algorithm 1 are considered. For clarity they are rewritten as

$$R'_{\overline{d_j}} \leftarrow R_0 \times M \mod N \tag{1}$$

$$R''_0 \leftarrow R_0'^2 \mod N. \tag{2}$$

If $\overline{d_j} = 1$, the memory R_0 is not updated in Eq. (1) and thus $R_0 = R'_0$. Therefore, the multiplication and squaring collide. Alternatively when $\overline{d_j} = 0$, $R_0 \neq R'_0$ and there is no collision. As a result, one bit of the secret exponent is revealed by sensing the collision.

Suppose N different messages are encrypted with the same exponent. The multiplication and squaring traces for the i-the message are denoted by m_x^i and s_x^i, respectively. The subscript x represents time. In order to find the collision, m_x^i and s_x^i are compared. More specifically, the correlation coefficient matrix $C_{x,y}$ is calculated as:

$$C_{x,y} = \mathcal{F}_j[m_x^j, s_y^j]. \tag{3}$$

Here, $\mathcal{F}_j$ is the correlation-coefficient operator defined as follows:

$$\mathcal{F}_j[t^j, s^j] := \frac{1}{N} \sum_{j=0}^{N-1} \frac{(t^j - \mathcal{E}_j[t^j]) \cdot (s^j - \mathcal{E}_j[s^j])}{\sqrt{(\mathcal{E}_j[t^j]^2 - \mathcal{E}_j[(t^j)^2]) \cdot (\mathcal{E}_j[s^j]^2 - \mathcal{E}_j[(s^j)^2])}}, \tag{4}$$

$$\mathcal{E}_j[t^j] := \frac{1}{N} \sum_{j=0}^{N-1} t^j. \tag{5}$$

If there is a collision, $C_{x,y}$ contains a non-zero value. Therefore, the collision can be found by looking at the matrix.

2.3   Single-Shot Collision Attacks by Hanley et al. [8]

Hanley et al. proposed a single-shot attack against various addition-chain algorithms. In the following description, we focus on the one for the multiply-always method.

The attack uses internal collision similarly to the one by Witteman et al. However, the correlation coefficient in Eq. (4) is meaningless when $N = 1$ i.e., under a single-shot attack[1]. Instead, the two time-domain traces m_x^0 and s_x^0 are directly compared. Hanley et al. presented two different ways to measure the similarity: the Euclidean distance and the time-domain correlation coefficient given by $\mathcal{F}_x[m_x^0, s_x^0]$.

Hanley et al. applied the attack to a software implementation and successfully recovered 99 % of the exponent bits. They also applied the attack to an FPGA implementation, however, the attempt was unsuccessful with the one with the multiply-always method. That is explained by the fact that (i) single-shot attacks are susceptible to SNR and (ii) SNR is usually low in FPGAs because of higher parallelism.

The attack uses multiple points of interest. That is the advantage of the attack over the clustering-based attacks [7]. Therefore, the multiply-always method can be defeated even if there is no first-order leakage [7]. In addition, the attack is advantageous to HCPA on the point the known message is not needed. In other words, the attack by Hanley et al. defeats the message-blinding countermeasure.

---

[1] Clavier et al. proposed another single-shot extension [15]. The purpose of the attack is to distinguish multiplication and squaring. There is a practical result on a software implementation.

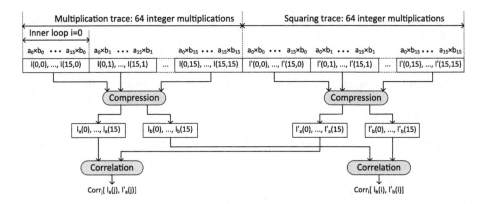

Fig. 1. Distinguisher

2.4   Proposed Extension of the Attack by Hanley et al.

An extension of the attack by Hanley et al. is described. The extension is on measuring the similarity between the traces. The idea is simple: the signal processing of BMA is applied to the traces before comparison[2].

Long-integer multiplication $A \times B$ is considered. A and B are composed of s words. They are denoted by $A = \{a_{s-1}, \cdots, a_0\}$ and $B = \{b_{s-1}, \cdots, b_0\}$ where a_j and b_i are words. The long-integer multiplication comprises generation of partial products $a_j \times b_i$. The leakage of $a_j \times b_i$ is denoted by $l(j, i)$.

The trace $l(j, i)$ is processed before comparison. Figure 1 illustrates the process. Firstly, $l(j, i)$ is compressed into s-dimensional vectors $l_a(j)$ and $l_b(i)$. They are defined as:

$$l_a(j) = \frac{1}{s} \sum_{i=0}^{s-1} l(j, i), \tag{6}$$

$$l_b(i) = \frac{1}{s} \sum_{j=0}^{s-1} l(j, i). \tag{7}$$

$l_a(j)$ and $l_b(i)$ are called the compressed vectors. By the compression, the effect of one operand is removed thereby SNR of another operand is improved. The compressed vectors $l_a(j)$ and $l_b(i)$ correlate to a_j and b_i, respectively.

Finally, the compressed vectors from multiplication and squaring traces are compared in the same manner as the original attack. The measured traces of multiplication and squaring are denoted by $l(j, i)$ and $l'(j, i)$, respectively. The corresponding compressed vectors are denoted by $l_a(j)$, $l'_a(j)$, $l_b(i)$, and $l'_b(i)$. If the time-domain correlation coefficient is used for measurement, they are expressed as:

---

[2] The method can be thought as a missing variant with "regular algorithm + unknown message" in the categorization by Bauer et al. [10].

$$\mathcal{F}_j[l_a(j), l'_a(j)] \in [-1, 1], \tag{8}$$

$$\mathcal{F}_i[l_b(i), l'_b(i)] \in [-1, 1]. \tag{9}$$

The correlation coefficients become high if there is collision.

In order to conduct the attack, the attacker needs to get $l(j, i)$ from a raw trace in the same manner as BMA and HCPA. Even if the prior knowledge is unavailable, the attacker can possibly reverse-engineer the points of $l(j, i)$ by analyzing the correlation matrix in Eq. (3). That is because the patterns on the matrix reflect the underlying long-integer multiplication algorithm. That is explained in Sect. 4.2 with experimental results. Note that in order to get a meaningful correlation matrix, the exponent blinding should be disabled. Such a requirement is satisfied in two cases. Firstly, the attacker with an open sample can possibly profile the device while disabling the countermeasure. Secondly, the same co-processor for modular exponentiation may be used for another purpose without the exponent blinding. One such example is signature verification in which no secret is involved.

3   Asymmetric Leakage

Difference between two operands of multipliers is discussed in (i) integer multiplier and (ii) long-integer multiplication (LIM) levels.

3.1   Asymmetry at Integer Multiplier Level

In the paper of BMA, Walter showed that two operands of a simple multiplier are symmetric in terms of side-channel leakage [12]. However, sophisticated multipliers can be asymmetric as described below.

The Booth recoding is a common technique for partial product generation (PPG)[3] [17]. The technique enables to reduce the total number of partial products thereby improving the performance of integer multiplication. Figure 2 shows a circuit for generating one partial product using the radix-4 Booth recoding. Firstly, the multiplicand A is expanded to $\{\overline{2A}, \overline{A}, 0, A, 2A\}$. The expansion is efficiently implemented using shifts and NOT gates. Then, one out of the five candidates is selected at the 5:1 selector. The selector output is the partial product. The selector is controlled by a 3-bit chunk of the multiplier namely $\{x_{i+1}, x_i, x_{i-1}\}$.

The circuit in Fig. 2 is asymmetric between operands. Therefore, asymmetric leakage is expected. Leakage from a 32-bit integer multiplier with the radix-4 Booth recoding is simulated. The multiplier is synthesized and post-synthesis logic simulation is conducted. While the logic simulation, the number of signal-transition events i.e., toggles is measured.

---

[3] Note that Walter and Samyde noticed that leakage from the Booth recoding is not the one by the Hamming-weight model [13]. However, the difference between operands was not discussed.

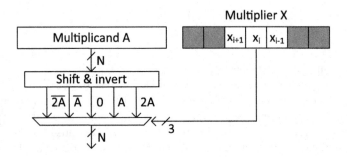

Fig. 2. A circuit for generating a partial product in the radix-4 Booth recoding

Two sets of test-vectors are used to drive the circuit. They are $c \times x_i$ and $x_i \times c$ where c and x_i are 32-bit integers. The test-vectors are designed to measure toggles from one operand by fixing another to a constant.

Histograms of the measured toggle counts are shown in Fig. 3. The black and white bars correspond to the two sets of test-vectors. The two sets show clearly different histograms. As shown in the histograms, more toggles are observed when the multiplicand is fixed. That means the multiplier port makes more toggles. The result is explained by an empirical fact that a selector signal has stronger effect on toggle counts. More specifically, toggles at the 3-bit control signal is amplified to N bits at the selector output.

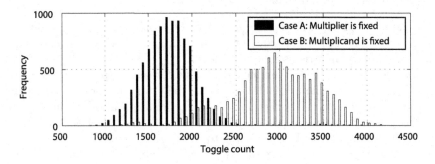

Fig. 3. Toggle counts of a 32-bit multiplier

3.2 Asymmetry at Long-Integer Multiplication Level

Difference of operands at the LIM level is discussed. There are many options at this level. The Montgomery multiplication with the coarsely integrated operand scanning (CIOS) shown in Algorithm 2 is considered.

The long integers are represented by $A = \{a_{s-1}, \cdots, a_0\}$ and $B = \{b_{s-1}, \cdots, b_0\}$ where a_j and b_i are words. The core operation is $a_j \times b_i$ at the line 4 of Algorithm 2 in which partial products are generated. LIM is commonly implemented with a circuit shown in Fig. 4. The circuit uses a multiply-and-accumulate

---

Algorithm 2. Coarsely Integrated Operand Scanning [16]

Input: Word $A = \{a_{s-1}, \cdots, a_0\}$ and $B = \{b_{s-1}, \cdots, b_0\}$
Output: Product t_i for $i \in [0, s-1]$
 1: **for** $i = 0$ to $s - 1$ **do**
 2:    $C \leftarrow 0$
 3:    **for** $j = 0$ to $s - 1$ **do**
 4:       $(C, S) \leftarrow t_j + a_j \times b_i + C$
 5:       $t_j \leftarrow S$
 6:    **end for**
 7:    $(C, S) \leftarrow t_s + C$
 8:    $t_s \leftarrow S$
 9:    $t_{s+1} \leftarrow C$
10:    $C \leftarrow 0$
11:    # Lines for the Montgomery reduction are not displayed for clarity.
12:    # See the literature [16] for the complete list.
13: **end for**
14: Return t_j

(MAC) unit. The words a_j and b_i are read from the memory and fed to the MAC unit via temporal registers labeled regA and regB.

Suppose regA and regB store the long integers A and B, respectively. Figure 4 also shows an operation sequence describing the contents of the registers. As shown in the table, regB is updated less frequently because the stored variables b_i is scanned at the outer loop in Algorithm 2. For s-word long integers, regA and regB are updated s^2 and s times, respectively.

CMOS circuits make data-dependent power consumption when their inputs are changed [2]. As a result, the operand scanned at the inner loop (i.e., A) has stronger leakage. This LIM-level asymmetry is verified through experiments in Sect. 4.

4 Experiments

Traces are captured from an FPGA implementation of the Montgomery multiplication. Firstly, the traces are analyzed using the attack by Witteman et al. Then, the single-shot attack in Sect. 2.4 is applied. The purpose of the experiment is twofold. Firstly, feasibility of the attack in Sect. 2.4 is verified. Secondly, the effects of the asymmetric leakage are examined.

4.1 Setup

A circuit implementing the 1024-bit Montgomery multiplication is examined. The circuit uses the MAC-based architecture in Fig. 4. The MAC unit has a 64-bit integer multiplier and thus the number of words $s = 16 = 1024/64$. The words are scanned with the CIOS method in Algorithm 2. The two operands to the integer multiplier can be swapped by an external signal in order to evaluate

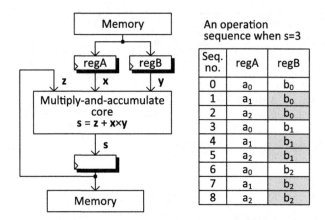

Fig. 4. A common circuit architecture for LIM

the asymmetry at the integer-multiplier level. The target circuit is implemented on Virtex-II Pro FPGA on SASEBO [20].

The FPGA is measured by putting a magnetic-field probe on the chip surface [18]. The probe is 0.1 mm in diameter. Traces are captured using an oscilloscope with the bandwidth of 12.5 GHz and the sampling rate of 25.0 GSa/s.

Test-vectors are designed to emulate RSA with the multiply-always method (see Algorithm 1). Firstly, 1024-bit random numbers x_k, y_k, and z_k are generated for $0 \leq k < 1000$. For each triplet (x_k, y_k, z_k), the Montgomery multiplication is called in five ways as summarized in Table 1. Note that the Montgomery multiplication is denoted by $\mathcal{M}(\cdot, \cdot)$ in the table.

The five Montgomery multiplications are denoted by (XY1), (XY2), (XX), (YY), and (ZZ). The Montgomery multiplication $\mathcal{M}(x_k, y_k)$, which corresponds to the multiplication in Algorithm 1, is conducted in (XY1) and (XY2). (XY1) and (XY2) are different in the order of operands to the 64-bit integer multiplier. See Table 1 for the operands in the LIM and integer-multiplier levels. The remaining Montgomery multiplications namely $\mathcal{M}(x_k, x_k)$, $\mathcal{M}(y_k, y_k)$, and $\mathcal{M}(z_k, z_k)$ correspond to the squaring in Algorithm 1.

The traces are examined in pair. Six pairs namely

$$\{(\text{XY1}), (\text{XY2})\} \times \{(\text{XX}), (\text{YY}), (\text{ZZ})\}.$$

are evaluated. The pairs are referred to as (i)-(vi) as summarized in Table 2. There are collisions in the pairs (i)-(iv). Colliding operands, both at the LIM and integer-multiplier levels, are also shown in Table 2. In (i) and (iii), there is a collision between the LIM-level operands scanned at the inner loop of Algorithm 2. On the other hand, the operands scanned at the outer loop collide in (ii) and (iv). At the integer-multiplier level, the multiplicands collide in (i) and (iv). Alternatively the multipliers collide in (ii) and (iii). There is no collision in (v) and (vi).

The pairs are compared under the multiple- and single-shot attacks in the following sections.

Table 1. Test-vectors of the Montgomery multiplication

Identifier	Operation	LIM level		Integer-multiplier level	
		Inner loop (a_j)	Outer loop (b_i)	multiplier	multiplicand
(XY1)	$\mathcal{M}(x_k, y_k)$	x_k	y_k	y_k	x_k
(XY2)	$\mathcal{M}(x_k, y_k)$	x_k	y_k	x_k	y_k
(XX)	$\mathcal{M}(x_k, x_k)$	x_k	x_k	x_k	x_k
(YY)	$\mathcal{M}(y_k, y_k)$	y_k	y_k	y_k	y_k
(ZZ)	$\mathcal{M}(z_k, z_k)$	z_k	z_k	z_k	z_k

Table 2. Examined pairs of traces

Identifier	Multiplication m_x^j	Squaring s_y^j	LIM-level collision	Integer-multiplier-level Collision
(i)	(XY1)	(XX)	inner (a_j)	multiplicand
(ii)	(XY1)	(YY)	outer (b_i)	multiplier
(iii)	(XY2)	(XX)	inner (a_j)	multiplier
(iv)	(XY2)	(YY)	outer (b_i)	multiplicand
(v)	(XY1)	(ZZ)	—	—
(vi)	(XY2)	(ZZ)	—	—

4.2   Multiple-Shot Leakage Using the Attack by Witteman et al.

As a preliminary experiment, the pairs of the traces are analyzed using the attack by Witteman et al. The correlation matrices in Eq. (3) are calculated for the pairs (i)-(iv) in Table 2. The matrices are visualized as bitmap images in Fig. 5.

The bitmap images show different patterns depending on the colliding operands at the LIM level. There are repeated slash lines on Fig. 5-(i) and -(iii) in which there are collisions at the inner loop. On the other hand, collision at the outer loop makes rectangle patterns as shown in Fig. 5-(ii).

The bitmap images also show the difference caused by the asymmetry at the integer-multiplier level. The multiplier (cf. the multiplicand) shows higher correlation as expected in Sect. 3.1. The slash lines are more clear in Fig. 5-(iii) compared to the ones in Fig. 5-(i). Similarly, the rectangle patterns are more distinct in Fig. 5-(ii).

The bitmap images are intuitive but unsuitable for quantitative comparison. More concrete comparison is conducted in the next section.

As described in Sect. 2.4, the attacker can get the points of interest for $l(j, i)$ by interpreting the bitmap images. The above-mentioned slash-line and rectangle patterns are commonly found in many implementations. The attacker can sample the clock cycles with high correlation for $l(j, i)$.

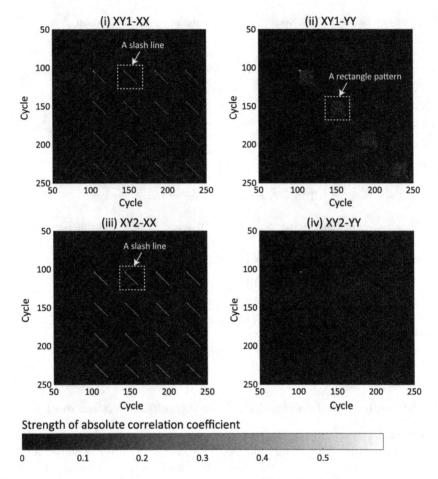

Fig. 5. Correlation coefficient matrices

4.3   Single-Shot Attack

The pairs of traces are analyzed with the method described in Sect. 2.4. The correlation coefficients are evaluated for the pairs (i)–(vi) in Table 2. The processing is chosen considering the patterns on the bitmap images. For the pairs (i) and (iii) with collisions at the inner loop, Eq. (8) is used. Alternatively, Eq. (9) is used for the pairs (ii) and (iv).

The results are shown as histograms of the correlation coefficients in Fig. 6. Figure 6-(i) to -(iv) correspond to the pairs (i)-(iv) in Table 2. The pairs (v) and (vi), that have no collision, are also shown for comparison.

In the real attack, the black and white histograms are not separated. Therefore, the attacker should set a threshold to make a decision. In this experiment, the measured correlation coefficients are split into upper and lower halves. In other words, median is used as a threshold. Finally, the rate of successful deci-

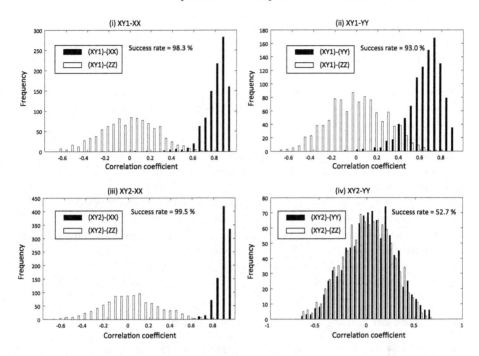

Fig. 6. Histograms of correlation coefficients

sion is calculated. The success rates are (i) 98.3 %, (ii) 93.0 %, (iii) 99.5 %, and (iv) 52.7 %, respectively. The pairs (iii) is the most distinguishable. This is the first successful single-shot collision attack of the multiply-always method on FPGA. In contrast, the attack is unsuccessful in (iv). The result show that operand order has a significant impact on the success rate of the attack.

5 Discussion

The experimental results indicate that the operand order has a considerable impact on side-channel leakage. The results are discussed from countermeasure and attack view points.

5.1 Leak Reduction by Designing Operand Order

The experimental results show that the operand order determines success and failure of the attack. Therefore, the amount of leakage can be reduced by appropriately designing the order of operands. The pair (iv) in Table 2 is the best option (see Fig. 6-(iv)). The operand order can be changed at almost no cost. In addition to the cost effectiveness, the proposed method can easily be combined with other conventional countermeasures (e.g., the randomized operand scanning [9,10]).

The target implementation discussed in the paper, the multiply-always method using the Montgomery multiplication with CIOS, is one of many possible designs. It is worth discussing how the operand order can be designed in other cases.

Firstly, the same idea can be easily extended to many other methods. That is because the causes of asymmetry, the partial-product generation and operand scanning, are essential in long-integer multiplication.

However, there is always an exception. A notable exception is the Finely Integrated Operand Scanning (FIOS) instead of CIOS [16]. In FIOS, the register containing the variable scanned at the outer cannot be kept while the outer loop. That is because another word-wise multiplication, needed for the Montgomery reduction, should be interleaved. As a result, the leakage from the operand scanned at outer loop is not necessarily smaller.

Alternatively, the operand order can be determined using the conventional toggle simulation. As shown in Sect. 3.1, the asymmetry at the integer multiplier can be simulated. The LIM-level asymmetry is not simulated in the paper, however, the frequency of register update can be covered by the toggle simulation.

5.2   Attack on Montgomery Powering Ladder

In contrast to the previous result, the asymmetric leakage can make some countermeasures ineffective. Algorithm 3 shows the Montgomery powering ladder (MPL). We focus on collisions between inputs[4].

In MPL, there is always collision between consecutive operations:

$$R_{\overline{a}} \leftarrow R_0 \times R_1 \quad \mathrm{mod}\ N, \tag{10}$$

$$R_a \leftarrow R_a \times R_a \quad \mathrm{mod}\ N. \tag{11}$$

Therefore, the presence of collision does not leak k_j. However, the colliding operand depends on k_j. If $k_j = a = 0$, the first operands in Eqs. (10) and (11) collide. On the other hand, the second operands collide if $k_j = a = 1$. If the attacker can distinguish the collisions at first and second operands, the secret parameter $k_j = a$ is revealed.

Interestingly, if Eqs. (10) and (11) are replaced with the following statements without changing the result of the algorithm, the attack is no longer effective:

$$R_{\overline{a}} \leftarrow R_a \times R_{\overline{a}} \quad \mathrm{mod}\ N, \tag{12}$$

$$R_a \leftarrow R_a \times R_a \quad \mathrm{mod}\ N. \tag{13}$$

The algorithm appear in the work by Hanley et al. [8]. Now, collision is always occurred at the first operand. Therefore, the colliding operand becomes independent of k_j. This is another example showing the importance of designing operand order.

---

[4] Hanley et al. considered more general cases considering a collision between input and output. However, the input-output collision was very weak in our setup.

Algorithm 3. Montgomery Powering Ladder

Input: Message M, scalar $k = (k_{t-1}, \cdots, k_0)_2$
Output: Ciphertext M^d
1: $R_0 \leftarrow 1; R_1 \leftarrow M$
2: **for** $j = t - 1$ downto 0 **do**
3:    $a \leftarrow k_j; \overline{a} \leftarrow 1 - a$
4:    $R_{\overline{a}} \leftarrow R_0 \times R_1 \mod N$
5:    $R_a \leftarrow R_a \times R_a \mod N$
6: **end for**
7: Return R_0

---

6 Conclusion

Two operands of multipliers are asymmetric in terms of side-channel leakage. The reason can be explained by asymmetries at arithmetic-circuit and micro-architecture levels. The leakage can be suppressed by appropriately designing the order of operands. On the other hand, some countermeasure can be defeated if the leakages from first and second operands are distinguishable.

Many problems are remained open. The attack using input-to-output collision is an interesting challenge. Another important open problem is on incomplete exponent recovery. The successful rate more than 99 % is clearly dangerous. The ideal goal is 50.0 %, however, it could possibly be relaxed.

Acknowledgement. The authors would like to thank the anonymous reviewers at COSADE 2015 for their valuable comments. The study was conducted as a part of the CREST Dependable VLSI Systems Project funded by the Japan Science and Technology Agency.

References

1. Kocher, P.C., Jaffe, J., Jun, B.: Differential power analysis. In: Wiener, M. (ed.) CRYPTO 1999. LNCS, vol. 1666, pp. 388–397. Springer, Heidelberg (1999)
2. Mangard, S., Oswald, E., Popp, T.: Power Analysis Attacks: Revealing the Secrets of Smart Cards. Springer-Verlag, New York (2007)
3. Coron, J.-S.: Resistance against differential power analysis for elliptic curve cryptosystems. In: Koç, Ç.K., Paar, C. (eds.) CHES 1999. LNCS, vol. 1717, p. 292. Springer, Heidelberg (1999)
4. Amiel, F., Feix, B., Tunstall, M., Whelan, C., Marnane, W.P.: Distinguishing multiplications from squaring operations. In: Avanzi, R.M., Keliher, L., Sica, F. (eds.) SAC 2008. LNCS, vol. 5381, pp. 346–360. Springer, Heidelberg (2009)
5. Homma, N., Miyamoto, A., Aoki, T., Satoh, A., Shamir, A.: Collision-based power analysis of modular exponentiation using chosen-message pairs. In: Oswald, E., Rohatgi, P. (eds.) CHES 2008. LNCS, vol. 5154, pp. 15–29. Springer, Heidelberg (2008)

6. Heyszl, J., Ibing, A., Mangard, S., De Santis, F., Sigl, G.: Clustering algorithms for non-profiled single-execution attacks on exponentiations. In: Francillon, A., Rohatgi, P. (eds.) CARDIS 2013. LNCS, vol. 8419, pp. 79–93. Springer, Heidelberg (2014)

7. Perin, G., Imbert, L., Torres, L., Maurine, P.: Attacking randomized exponentiations using unsupervised learning. In: Prouff, E. (ed.) COSADE 2014. LNCS, vol. 8622, pp. 144–160. Springer, Heidelberg (2014)

8. Hanley, N., Kim, H.,Tunstall, M.: Exploiting collisions in addition chain-based exponentiation algorithms using a single trace. Cryptography ePrint Archive: Report 2012/485. http://eprint.iacr.org/2012/485

9. Clavier, C., Feix, B., Gagnerot, G., Roussellet, M., Verneuil, V.: Horizontal correlation analysis on exponentiation. In: Soriano, M., Qing, S., López, J. (eds.) ICICS 2010. LNCS, vol. 6476, pp. 46–61. Springer, Heidelberg (2010)

10. Bauer, A., Jaulmes, E., Prouff, E., Wild, J.: Horizontal and vertical side-channel attacks against secure RSA implementations. In: Dawson, E. (ed.) CT-RSA 2013. LNCS, vol. 7779, pp. 1–17. Springer, Heidelberg (2013)

11. Bajard, J.-C., Imbert, L., Liardet, P.-Y., Teglia, Y.: Leak resistant arithmetic. In: Joye, M., Quisquater, J.-J. (eds.) CHES 2004. LNCS, vol. 3156, pp. 62–75. Springer, Heidelberg (2004)

12. Walter, C.D.: Sliding windows succumbs to Big Mac attack. In: Koç, Ç.K., Naccache, D., Paar, C. (eds.) CHES 2001. LNCS, vol. 2162, p. 286. Springer, Heidelberg (2001)

13. Walter, C.D.,Samyde, D.: Data Dependent Power Use in Multipliers. In: 17th IEEE Symposium on Computer Arithmetic (ARITH 2005)

14. Witteman, M.F., van Woudenberg, J.G.J., Menarini, F.: Defeating RSA multiply-always and message blinding countermeasures. In: Kiayias, A. (ed.) CT-RSA 2011. LNCS, vol. 6558, pp. 77–88. Springer, Heidelberg (2011)

15. Clavier, C., Feix, B., Gagnerot, G., Giraud, C., Roussellet, M., Verneuil, V.: ROSETTA for single trace analysis. In: Galbraith, S., Nandi, M. (eds.) INDOCRYPT 2012. LNCS, vol. 7668, pp. 140–155. Springer, Heidelberg (2012)

16. Koç, C.K., Acar, T., Kaliski Jr, B.S.: Analyzing and comparing montgomery multiplication algorithms. Micro, IEEE **16**(3), 26–33 (1996)

17. Koren, I.: Computer Arithmetic Algorithms, 2nd edn. A K Peters, CRC Press, Boston, Boca Raton (2001)

18. Sugawara, T., Suzuki, D., Saeki, M., Shiozaki, M., Fujino, T.: On measurable side-channel leaks inside ASIC design primitives. In: Bertoni, G., Coron, J.-S. (eds.) CHES 2013. LNCS, vol. 8086, pp. 159–178. Springer, Heidelberg (2013)

19. Okeya, K., Sakurai, K.: A second-order DPA attack breaks a window-method based countermeasure against side channel attacks. In: Chan, A.H., Gligor, V.D. (eds.) ISC 2002. LNCS, vol. 2433, p. 389. Springer, Heidelberg (2002)

20. AIST, Side-Channel Attack Standard Evaluation Board. http://www.risec.aist.go.jp/project/sasebo/

FPGA Countermeasures

Evaluating the Duplication of Dual-Rail Precharge Logics on FPGAs

Alexander Wild$^{(\boxtimes)}$, Amir Moradi, and Tim Güneysu

Horst Gortz Institute for IT-Security, Ruhr-Universitat Bochum, Bochum, Germany
{alexander.wild,amir.moradi,tim.gueneysu}@rub.de

Abstract. Power-equalization schemes for digital circuits aim to harden cryptographic designs against power analysis attacks. With respect to dual-rail logics most of these schemes have originally been designed for ASIC platforms, but much efforts have been spent to map them to FPGAs as well. A particular challenge is here to apply those schemes to the predefined logic structures of FPGAs (i.e., slices, LUTs, FFs, and routing switch boxes) for which special tools are required. Due to the absence of such routing tools Yu and Schaumont presented the idea of duplicating (i.e., dualizing) a fully-placed-and-routed dual-rail precharge circuit with equivalent routing structures on an FPGA. They adopted such architecture from WDDL providing the Double WDDL (DWDDL) scheme.

In this work we show that this general technique – regardless of the underlying dual-rail logic – is incapable to properly prevent side-channel leakages. Besides theoretical investigations on this issue we present practical evaluations on a Spartan-6 FPGA to demonstrate the flaws in such an approach. In detail, we consider an AES-128 encryption module realized by three dual-rail precharge logic styles as a case study and show that none of those schemes can provide the desired level of protection.

1 Introduction

Side-Channel Analysis (SCA) is of major challenges for secure-hardware designers. The most popular techniques to harden a design are hiding, masking and leakage resilient architectures. The goal of power equalization schemes, which are a part of the hiding category, is to equalize the power consumption of a cryptographic circuit independent of the processed data. In hardware devices such schemes often follow the Dual-rail Precharge Logic (DPL) concept like SABL [21], WDDL [22], DRSL [6], MDPL [18] and iMDPL [17]. Most of the dual-rail schemes have been developed to be implemented in ASICs. The fixed architecture as well as limited wire routings on FPGAs does not allow a straight-forward porting of those schemes.

1.1 Related Work

During the last years, much effort was put in the direction of bringing the DPL concept to FPGAs. Nassar et al. [16] introduced Balanced Cell-based Dual-rail

© Springer International Publishing Switzerland 2015
S. Mangard and A.Y. Poschmann (Eds.): COSADE 2015, LNCS 9064, pp. 81–94, 2015.
DOI: 10.1007/978-3-319-21476-4_6

Logic (BCDL) that connects a global precharge signal with all gate inputs to a rendezvous box which is placed in front of each gate. The rendezvous box triggers an evaluation signal for the corresponding gate in case of stable input signals. This work did not consider the routing of the *true* and *false* networks. So the wire capacities of their dual-rail networks will be different. He et al. followed an approximately similar approach [8]. In their work the evaluation of a gate is triggered by two global signals that are directly connected to the gates. Lomné et al. [11] followed a triple-rail approach where an artificially delayed asynchronous control signal is used. The additional delay guarantees the latest arriving of the control signal which triggers the gate evaluation and therefore prevents the gate from early propagation (EP).

In a more recent work [9] He et al. applied duplication while minimizing the area overhead. They duplicated a fully placed and routed circuit to realize the dualization. The original and duplicated circuit can be interleaved, which may leads to routing conflicts. Hence, a method has been developed to detect and correct these routing conflicts. As a result, the circuits are differently routed at some points which lead to side-channel information leakage.

In [19], Sauvage et al. evaluated different placement strategies to support the router in finding routes for the dual networks with minimal delay differences. The strategies they followed were first, placing the components of the original circuit as close as possible together as well as the components of the dual circuit and second, placing the related components of the original and dual circuit as close as possible together. It turned out that the placement did not have the desired impact on the routing.

Bhasin et al. introduced in their work [3] an improved version of WDDL called DPL-noEE. Since it has been shown that WDDL suffers from the early propagation effect [20], DPL-noEE connects the signals of the true and false network to the same gate and evaluates the gate output with the arrival of the last incoming signal. According to the authors' report, compared to WDDL the leakage is approximately halved by means of DPL-noEE. It has been pointed out in [15] that DPL-noEE just solves the early propagation effect in the evaluation phase of the circuit but not in the precharge phase. They introduced a logic style called AWDDL that solved this problem. An AWDDL gate switches to precharge state when the last input signal turns to precharge. Additionally, a customized router was developed that tries to find routes with minimal delay differences for the true and false network by moving the routing process into a Satisfiability (SAT) solvable problem. It is also noted by the authors that the router could minimize the leakage of the circuit but did not completely remove it due to the nonexistence of perfectly-identical dual-rail routes.

To deal with routing imbalances, Yu and Schaumont had formerly proposed to duplicate a fully-placed-and-routed WDDL circuit [24] (known as DWDDL). As a result, two equivalently routed WDDL circuits with swapped true and false networks (dualized) were placed on the same device. With investment of doubled resources, the leakage of a WDDL circuit was drastically reduced.

1.2   Motivation and Contribution

Power-equalization schemes place a circuit C and the dual of the circuit $\overline{C}$ on the same device. Ideally, the total power consumption of both circuits cancel out the data dependency in the power traces. This idea implies that the logic gates of C and $\overline{C}$ have to be synchronized and switch at the same point in time. This synchronization can be achieved by a global or asynchronous control signal connected to the gates like in [8,11,16] or with the arrival of the input signals at the gate like in [3,15,22]. To use the input signals as a trigger for the evaluation, the signals of the dualized circuit $\overline{C}$ must show exactly the same delays as their pendant signals in C. Beside the synchronization aspect, the signal delays are strongly correlated with the capacity of the used wires so that even in a control signal synchronized circuit the signal delays of both circuits shall be equivalent to minimize the data-dependent side-channel leakage. The task of placing two circuits with the same signal delays is hard to achieve in FPGAs due to the static routing structure. Some logic styles, e.g., the most popular one WDDL, require the interconnection of both C and $\overline{C}$ circuits. In such cases the task to place and route the logic gates in such a way that the routing delays of coupled signals are equivalent is challenging. In [15] this task was addressed with a custom router based on a SAT solver.

As stated, in DWDDL [24] the fully-routed-and-interconnected original circuit $(C, \overline{C})$ is additionally placed with inverted logic on the same device $(\overline{C'}, C')$. The cloning process was performed in a way that the routing information is transferred to the cloned circuit. According to the report of the Xilinx design tools, the signal delays of the original circuits $(C, \overline{C})$ and the cloned circuits $(\overline{C'}, C')$ are equivalent. So this method turned out to be the best way to implement two circuits of equivalent routing delays without any routing restrictions or any custom routers. The drawback of the technique is clearly the high resource overhead.

In this work we thoroughly investigate the duplication scheme of [24]. Since the early propagation issue of WDDL makes it still vulnerable to the state-of-the-art attacks, we consider its successors DPL-noEE and AWDDL as well to examine the benefit of the duplication. We show that even in case of AWDDL, where early propagation at both phases is avoided, applying the duplication does not prevent data-dependent time of evaluation. By means of a Spartan-6 evaluation platform (SAKURA-G [1]) we provide practical evidences to our findings that a DWDDL circuit and the equivalent ones realized by DPL-noEE as well as AWDDL still have leakage.

2   Logic Styles

Each of WDDL, DPL-noEE and AWDDL consists of gates with two outputs, O_t and O_f. In the precharge phase both outputs have the same value ($O_t = 0, O_f = 0$), while in the evaluation phase only one output changes its state so that exactly one transition per evaluation phase is guaranteed. O_t presents the true value while O_f the false pendant. The true outputs form a network we address as *true network* (respectively *false network* made by the false outputs). In case of a negative gate, e.g., NAND or NOR, the corresponding non-negative gate (resp. AND or OR) is instantiated with switched output signals. Clearly, an inverter gate is realized by a connection switch swapping the dual rails. Hence such logic styles form two logical circuits that are interconnected. We further refer to this dual-rail circuit as the original circuit $(C, \overline{C})$. Following the duplication scheme of [24], we clone the original circuit, invert the logic and place it at a different location on the FPGA. The circuit made by this process is denoted as the duplicated circuit $(\overline{C'}, C')$. Below we shortly recall the specification of each of our considered logic styles.

WDDL is one of the most common DPL styles and mainly designed for ASICs. In WDDL only AND/NAND and OR/NOR gates are allowed. An XOR/XNOR gate is constructed by two AND/NAND and one OR/NOR WDDL gates. As stated, our evaluation platform is a SAKRURA-G where a Xilinx Spartan-6 FPGA is plugged that is equipped with 6-to-2 LUTs. This gives the advantage to realize each WDDL gate by one LUT. The building blocks of WDDL – with respect to 6-to-2 LUTs – can be seen in Fig. 1(a).

DPL-noEE is unofficially the successor of WDDL. As stated in [20] and in [3], a WDDL gate evaluates its output at different points in time depending on the input data. For example, O_t of a WDDL OR gate is derived from two signals of the true network. Regardless of the other input, O_t goes high once one of its inputs of the true network goes high. This phenomena is known as early propagation effect and causes data-dependent power consumption and hence side-channel leakage.

This issue is addressed by the DPL-noEE logic style. As shown in Fig. 1(b), O_t of a DPL-noEE OR gate depends on all four input signals, so that the gate evaluates when all signals are available. Due to the consideration of the true and false signals in each gate – contrary to WDDL – a DPL-noEE XOR/XNOR gate can be constructed by a single LUT. This results in less LUT utilization particularly in designs with high number of XOR gates.

The authors of [15] noted in their work that DPL-noEE prevents the early propagation at the beginning of the evaluation phase. It is shown that the output of DPL-noEE gates fall back into precharge as soon as one of the input signals changes; hence the propagation in the precharge phase is faster. Although such an issue is not data dependent, it has been shown in [15] that the amount of power consumption (resp. amount of side-channel leakage) at the precharge phase is higher than that of the evaluation phase.

AWDDL emulates an SR-Latch inside each gate by looping the gate output to its input. Hence an AWDDL gate does not change its state until all input signals are in the evaluation or all in the precharge phase. Figure 1(c) shows the internal structure of AWDDL gates. In comparison to WDDL and DPL-noEE,

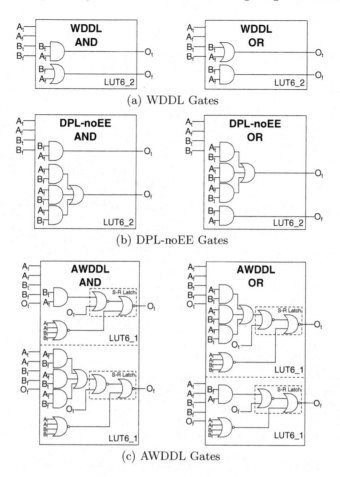

(a) WDDL Gates

(b) DPL-noEE Gates

(c) AWDDL Gates

Fig. 1. Logic style gates

one AWDDL gate should be realized by two 6-to-1 LUTs. In fact, a true 6-to-2 LUT would suffice to make both outputs of a gate. However, as shown by Fig. 2 the 6-to-2 LUT available in Xilinx FPGAs are made of two multiplexed 5-to-1 LUTs [23].

The necessity of employing two 6-to-1 LUTs per AWDDL gate as well as the gate loop paths increase the resource utilization and routing complexity compared to DPL-noEE. The authors of [15] have proposed to place both LUTs of each gate at the same slice to minimize the delay difference between the true and false outputs. Further, a customized router have been developed which tries to find equivalent routes for the true and false signals. The authors reported that the router improves the result but does not fully avoid the leakage associated to the signal delay differences. Table 1 summarizes the properties of the aforementioned logic styles.

Table 1. Properties of the in this work underlying logic style.

Logic style	Native gates	LUT/Gate	Address routing	EP
WDDL	AND, OR	1		in Eval. &Prech
DPL-noEE	AND, OR, XOR	1		in Prech
AWDDL	AND, OR, XOR	2	(✓)	-

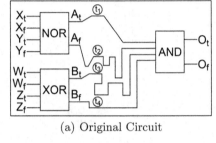

(a) Original Circuit

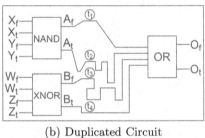

(b) Duplicated Circuit

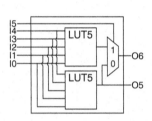

Fig. 2. 6-to-2 LUT

Fig. 3. Example of a fully routed circuit and its duplication.

3   Duplicating Circuits

This section deals with the idea proposed in [24] to duplicate a dual-rail circuit. Figure 3 shows an overview of an exemplary circuit and its duplicated counterpart. Due to the copied structure the signal routings and corresponding delays $t_1 \ldots t_4$ are transferred from the original to the duplicated circuit and just the logic gates are replaced. Attended by the gate replacement, the true and false networks are swapped in the duplicated circuit. Thus, the true network of the original circuit and the false network of the duplicated circuit (respectively the false network of the original circuit and the true network of the duplicated circuit) are equivalently routed that hence results in an overall design with very balanced true and false networks to minimize the leakage caused by different wire capacitances.

As stated before, DPL-noEE and AWDDL avoid the early propagation issue in contrast to WDDL. Along the same lines, the term *data-dependent time of evaluation* is defined as the cases when the gate evaluates its output at different time instances depending on its input values [12]. The duplication concept proposed in [24] (DWDDL) aims at mainly avoiding such a data-dependent time of evaluation caused by difference in routing of dual-rail signals. In the following we show that such a scheme cannot avoid data-dependent time of evaluation even if the underlying logic style prevents the early propagation.

3.1 Data-Dependent Time of Evaluation

In order to explain the concept we focus on the example given in Fig. 3. We just consider the start of the evaluation phase of an AND gate in both original and duplicated circuit (O_t and O_f in Fig. 3). Further, due to the early propagation issue of WDDL, we suppose that the gates in these circuits are realized by DPL-noEE or AWDDL (there is no difference in this example since both avoid early evaluation).

As marked in Fig. 3, we denote the delay of the input signals of the considered AND gate as t_1 to t_4, which stay the same for the corresponding OR gate in the duplicated circuit. We should highlight that no customized router is employed to route the signals in the original circuit. Indeed, the idea of duplication [24] is to avoid such a necessity. Therefore, the signal delays t_1 to t_4 can have any arbitrary value, and there is no guarantee to keep them the same or even approximately the same.

We first suppose that $t_1 > t_2 > t_3 > t_4$, and draw the timing diagram of the output signals (O_t and O_f) of both original and duplicated circuits for all four possible input values, e.g., 11, 10, 01, and 00. The corresponding timing diagram is shown by Fig. 4(a), where the black waveform belongs to the AND gate of original circuit (Fig. 3(a)) and the red waveform to its complementary OR gate of the duplicated circuit (Fig. 3(b)). Under such a condition the AND gate evaluates when the last input signal arrives. In case of 11 and 10 the output is evaluated at t_1 (when A_t arrives). For other input cases 01 and 00 t_2 defines the evaluation time (when A_f arrives). The duplicated OR gate shows a complementary behavior for the time of evaluation. That is, independent of the input value one gate always evaluates at t_1 and the other one at t_2. Hence, the overall power consumption is then ideally independent of the given input value. It should be noted that exchanging the values of t_1 by t_2 and/or t_3 by t_4 as well as (t_1, t_2) by (t_3, t_4) (i.e., providing another condition e.g., $t_3 > t_4 > t_2 > t_1$) has no effect on the shown balanced behavior.

Figure 4(b) gives the waveforms under a different condition $t_1 > t_3 > t_2 > t_4$. For the given condition the gates evaluate the outputs at t_1, t_2 or t_3. In case of 11 and 00, t_1 and t_2 defines the evaluation time, while for 01 and 10 the evaluation time depends on t_1 and t_3. In other words, one gate either in the original circuit or in the duplicated one evaluates at t_1, but the other gate evaluates at t_2 or at t_3 depending on the input value. This clearly shows a data-dependent time of evaluation and should lead to a leakage exploitable by an attack. Again we should note that exchanging t_1 by t_3 and/or t_2 by t_4 as well as (t_1, t_3) by (t_2, t_4) does not show any difference on the presented data-dependent time of evaluation.

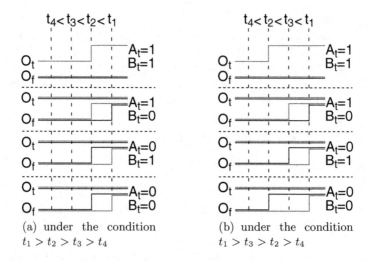

(a) under the condition
$t_1 > t_2 > t_3 > t_4$

(b) under the condition
$t_1 > t_3 > t_2 > t_4$

Fig. 4. Timing diagrams of the circuit of Fig. 3.

Remarks: In case of DWDDL the situation is much worse. That is, for every condition (except $t_1 = t_3$, $t_2 = t_4$) there is a data-dependent time of evaluation due to the early propagation of WDDL. The same issue holds true at the start of the precharge phase for all three considered logic styles. As a result, since in the certain conditions the duplication can be beneficial, we expect the duplication scheme to reduce but not fully avoid the leakage.

3.2   Duplication Tool

As our target platform is a Spartan-6 FPGA, we developed a tool to realize the duplication procedure. The tool clones the structure of the original circuit (in Xilinx Spartan-6 netlist format) and changes the LUTs content. It is indeed based on the same concept as the one introduced in [24]. Below a detailed description of the developed tool is given.

Xilinx netlists are stored in a proprietary file format (NCD). For low-level access Xilinx provides a tool to convert the proprietary netlist file format into (and back from) a human-readable format called Xilinx Design Language (XDL). This tool offers the ability to access and manipulate a fully-routed circuit on netlist level. Note that the XDL structure may change between FPGA families caused by technological progress in FPGA structure. The duplication tool was written with the help of RapidSmith [10] which is a Java-based Application Programming Interface (API) to read, write and edit XDL files. The reader interested in detailed information of the XDL file format is referred to [2].

The XDL file format is organized in instances and nets. An instance is an instantiated component on the device, e.g., a slice. The configuration of an instance is given by its attributes. These attributes also contain the LUT content. By changing the LUT content of an instance, the logic gates of a circuit can easily be changed. In case of the circuit duplication, an AND gate of the original circuit needs to be converted to an OR gate in the duplicated circuit, a NAND to a NOR and an XOR to an XNOR gate and vice versa. The physical location of an instance is defined in the instance header. Components of the Xilinx FPGAs are organized in a grid and can be identified by their X and Y coordinates. Hence, cloning a slice instance of the original circuit with manipulated LUT and placement coordinates adds a complementary-behaving instance to the design.

The routing of the signals is organized in a quite similar manner. The nets are routed via switch boxes that are able to interconnect different wires and are identifiable via their X and Y coordinates. A switch matrix configuration is called Programmable Interconnect Point (PIP). Hence, all PIPs used to route a net are written to the configuration part of that net. Copying the net with the edited PIP coordinates adds a net which is equivalently routed to a design. So, this process is performed for all nets of the original circuit. As an exception, the circuit I/O signals have to be handled differently. These nets cannot and are not needed to be duplicated. It is assumed, that the I/O signals (related to e.g., plaintext and ciphertext) do not leak any information. To route the I/O signals of the duplicated circuit, the standard Xilinx ISE routing tool cannot be used. In some cases it changes the already-routed nets, which may destroy the symmetry between the original and duplicated circuits. To route the I/O nets, the FPGA Editor which is also a part of the Xilinx ISE design suite offers the possibility to only route the specified signals. It is an adequate scenario for the low number of I/O signals.

To guarantee that the physical area, at which the cloned circuit should be placed, is unused by the remaining design, the Xilinx ISE synthesizer can be parametrized with the *prohibit* constraint to avoid placing any instance in that specified area. Another constraint we used is *area_group* to keep the original circuit in a specified area as well. We also made sure that no routing resources of the prohibited area is used by other logics.

4 Analysis

We implemented three AES-128 encryption cores using WDDL, DPL-noEE, and AWDDL gates. Following the scenario explained in Sect. 3.2, each of these cores is duplicated to realize the DWDDL, DDPL-noEE and DAWDDL AES cores. Hence, in total we analyze six full AES encryption cores. In general, for a fair comparison the best would be to keep the placement and routing of all these cores the same. However, WDDL has no defined native XOR gate; so a DPL-noEE circuit cannot be converted to a corresponding WDDL one. Also, AWDDL requires feedback loops as well as two LUTs per gate, hence converting an AWDDL circuit to a corresponding DPL-noEE one would be not fair with respect to resource utilization. Therefore, an identical placement and routing for these three logic styles is not possible.

For the AES core we implemented a round-based architecture with a composite field Sbox of [5] (see Fig. 5). Due to the known issue of register cells in dual-rail logics [14], we followed the master-slave fashion for the registers, i.e., two register stages in each rail. Therefore, every cipher round is operated in two clock cycles: one for the precharge and the other one for the evaluation. Indeed, in an interleaved fashion one of the round register stages holds the AES state and the other one the precharge 0 value.

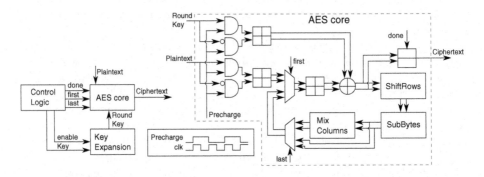

Fig. 5. Architecture of the full AES encryption with key expansion and control logic.

As we do not target any template attacks and ignore the leakage solely associated to the process on the key, the key expansion unit is implemented normally without using any secure logic style. As shown in Fig. 5, plaintext and round key (the output of the key expansion) are converted to a dual-rail precharge form and stored in dual-rail master-slave registers. For the sake of simplicity, we have not shown the key expansion and the control unit, which we kept equal regardless of the used logic style. The area which is marked (by a red dashed line) in Fig. 5 is the only part of the circuit that is implemented by WDDL, DPL-noEE, or AWDDL. Further, only this area is duplicated to realize DWDDL, DDPL-noEE, and DAWDDL circuits. The resource consumption of each core can be seen in Table 2.

Table 2. Overview of the implemented AES cores.

Logic style	Utilized resources		
	Slice	LUT	FF
WDDL	3,214	8,154	1,672
DWDDL	6,428	16,308	3,344
DPL-noEE	3,712	3,834	1,672
DDPL-noEE	7,424	7,668	3,344
AWDDL	3,724	7,146	1,672
DAWDDL	7,448	14,292	3,344
Control Logic	15	30	20
Key Expansion	83	307	132

4.1 Side-Channel Evaluation

For the practical evaluations we used SAKURA-G [1] as a side-channel evaluation board which is equipped with a Xilinx Spartan-6 FPGA. The power traces have been collected by monitoring the voltage drop over a $1\,\Omega$ resistor at Vdd path. The target FPGA operated at a frequency of $3\,\mathrm{MHz}$; power traces have been obtained by means of a digital oscilloscope at a sampling rate of $1\,\mathrm{GS/s}$. We also made use of the amplifier embedded on the SAKURA-G board and limited the acquisition bandwidth to $20\,\mathrm{MHz}$ to achieve high quality signals.

As the evaluation metric we used the leakage assessment methodology (t-test) presented in [7]. In a *non-specific* t-test (known also as *fix vs. random* test) a fix input (here plaintext) is selected, and during the measurements the fix or a random input is given to the target device[1]. The traces are categorized into two groups based on the associated fix or random input. Then, the Welch's (two-tailed) t-test is computed based on the mean and variance of each group of the traces (at each sample point independently). The outcome gives a level of confidence to reject a hypothesis as the traces of these two groups are drawn from the same population. If so (i.e., $|t| > 4.5$), it can be confidently concluded that a first-order leakage can be exploited from the device under evaluation.

Such a fix vs. random test is useful particularly to evaluate masked implementations. For instance, the same scheme has been used to examine the leakage of a higher-order attack resistant implementation of KATAN block cipher in [4]. However, in case of our designs where no masking is involved we cannot easily apply such a test. That is because the plaintext, which is not masked, is sent during the communication (between the FPGAs of the SAKURA-G) and processed by the target FPGA. Therefore, regardless of the protection that the AES core provides the leakage associated to the plaintext is observable at every sample point of the traces[2].

As a solution, a *semi fix* vs. random test can be performed. In such a scenario, a set of plaintexts are selected, all of which lead to the partially same intermediate value. We have precomputed 1024 plaintexts, in such a way that the first 64 bits of the cipher state at the start of the round 5 of the AES encryption are 0 (a similar scheme has been introduced in [7]). During the measurements a random plaintext or one of the precomputed plaintexts (again randomly) is taken. The rest of the evaluation stays the same as that of a fix vs. random test.

A sample trace of the WDDL design is shown at the top of Fig. 6. The first high peaks are related to the conversion and synchronization of the plaintext and the roundkey to the dual-rail precharge form (see Fig. 5). The trace of the other designs – particularly the duplicated ones – look like the same but with higher peak-to-peak value. For each of the six designs we collected $1\,000\,000$ traces following the scenario explained above. It means that approximately $500\,000$ traces with random plaintext and the rest with a randomly-chosen plaintext amongst the 1024 precomputed ones. The result of the tests on all six designs are shown by Fig. 6.

---

[1] Note that such a selection should also be randomized.

[2] One reason is also related to the static leakage [13].

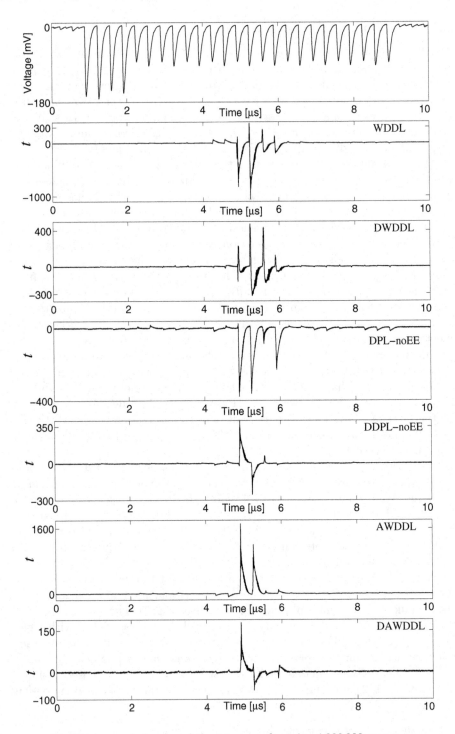

Fig. 6. A sample trace and the t-test results using $1\,000\,000$ traces.

It can be seen that none of the designs is able to avoid the leakage, and the tests strongly confirm the existence of a leakage. However, as we expected – stated in Sect. 3.1 – the duplicated designs can reduce the leakage, but due to the flaw (i.e., data-dependent time of evaluation for certain conditions) it cannot be prevented. It is noteworthy that the evaluation scheme which we performed is a "leakage assessment methodology" on one of the middle rounds of the cipher. Although we have not performed any key-recovery attack, and have not provided any information about the simplicity/hardness of such an attack, the result of the presented tests indicate the failure of the underlying design methodologies to prevent the leakages.

5   Conclusion

Regardless of its significant area overhead, duplicating a dual-rail precharge circuit (DWDDL) was considered as the only scheme that can be used for power equalization on FPGAs. In this work we have shown that this scheme is not flaw-less. By an extensive evaluation we found situations where the time of evaluation (of the gates in a double dual-rail precharge circuit) still depends on the input values. Such a data-dependent time of evaluation is caused by the difference between the signal delay of the gate inputs that cannot be avoided. Our theory is supported by practical analysis that we conducted on a Xilinx Spartan-6 FPGA. Although DPL-noEE and AWDDL avoid the well-known early propagation issue of WDDL, we have shown that still none of them can be considered as a potential solution to prevent the side-channel leakage.

Acknowledgment. This work was partially funded by the European Horizon 2020 project SAFEcrypto (grant no. 644729), German Research Foundation (DFG), and DFG Research Training Group GRK 1817/1.

References

1. Side-channel AttacK User Reference Architecture. http://satoh.cs.uec.ac.jp/SAK URA/index.html
2. Beckhoff, C., Koch, D., Tørresen, J.: The Xilinx Design Language (XDL): tutorial and use cases. In: ReCoSoC 2011, pp. 1–8. IEEE (2011)
3. Bhasin, S., Guilley, S., Flament, F., Selmane, N., Danger, J.: Countering early evaluation: an approach towards robust dual-rail precharge logic. In: WESS 2010, pp. 6. ACM (2010)
4. Bilgin, B., Gierlichs, B., Nikova, S., Nikov, V., Rijmen, V.: Higher-order threshold implementations. In: Sarkar, P., Iwata, T. (eds.) ASIACRYPT 2014, Part II. LNCS, vol. 8874, pp. 326–343. Springer, Heidelberg (2014)
5. Canright, D.: A very compact s-box for AES. In: Rao, J.R., Sunar, B. (eds.) CHES 2005. LNCS, vol. 3659, pp. 441–455. Springer, Heidelberg (2005)
6. Chen, Z., Zhou, Y.: Dual-rail random switching logic: a countermeasure to reduce side channel leakage. In: Goubin, L., Matsui, M. (eds.) CHES 2006. LNCS, vol. 4249, pp. 242–254. Springer, Heidelberg (2006)

7. Goodwill, G., Jun, B., Jaffe, J., Rohatgi, P.: A testing methodology for side-channel resistance validation. In: NIST Non-Invasive Attack Testing Workshop (2011)
8. He, W., de la Torre, E., Riesgo, T.: A precharge-absorbed DPL logic for reducing early propagation effects on FPGA implementations. In: ReConFig 2011, pp. 217–222. IEEE Computer Society (2011)
9. He, W., Otero, A., de la Torre, E., Riesgo, T.: Automatic generation of identical routing pairs for FPGA implemented DPL logic. In: ReConFig 2012, pp. 1–6. IEEE Computer Society (2012)
10. Lavin, C., Padilla, M., Lamprecht, J., Lundrigan, P., Nelson, B., Hutchings, B., Wirthlin, M.: RapidSmith - A Library for Low-level Manipulation of Partially Placed-and-Routed FPGA Designs. Technical report, Brigham Young University, September 2012
11. Lomné, V., Maurine, P., Torres, L., Robert, M., Soares, R., Calazans, N.: Evaluation on FPGA of triple rail logic robustness against DPA and DEMA. In: DATE 2009, pp. 634–639. IEEE Computer Society (2009)
12. Mangard, S., Oswald, E., Popp, T.: Power Analysis Attacks: Revealing the Secrets of Smart Cards. Springer, Heidelberg (2007)
13. Moradi, A.: Side-channel leakage through static power. In: Batina, L., Robshaw, M. (eds.) CHES 2014. LNCS, vol. 8731, pp. 562–579. Springer, Heidelberg (2014)
14. Moradi, A., Eisenbarth, T., Poschmann, A., Paar, C.: Power analysis of single-rail storage elements as used in MDPL. In: Lee, D., Hong, S. (eds.) ICISC 2009. LNCS, vol. 5984, pp. 146–160. Springer, Heidelberg (2010)
15. Moradi, A., Immler, V.: Early propagation and imbalanced routing, how to diminish in FPGAs. In: Batina, L., Robshaw, M. (eds.) CHES 2014. LNCS, vol. 8731, pp. 598–615. Springer, Heidelberg (2014)
16. Nassar, M., Bhasin, S., Danger, J., Duc, G., Guilley, S.: BCDL: a high speed balanced DPL for FPGA with global precharge and no early evaluation. In: DATE 2010, pp. 849–854. IEEE Computer Society (2010)
17. Popp, T., Kirschbaum, M., Zefferer, T., Mangard, S.: Evaluation of the masked logic style MDPL on a prototype chip. In: Paillier, P., Verbauwhede, I. (eds.) CHES 2007. LNCS, vol. 4727, pp. 81–94. Springer, Heidelberg (2007)
18. Popp, T., Mangard, S.: Masked dual-rail pre-charge logic: DPA-resistance without routing constraints. In: Rao, J.R., Sunar, B. (eds.) CHES 2005. LNCS, vol. 3659, pp. 172–186. Springer, Heidelberg (2005)
19. Sauvage, L., Nassar, M., Guilley, S., Flament, F., Danger, J., Mathieu, Y.: DPL on stratix II FPGA: what to expect?. In: ReConFig 2009, pp. 243–248. IEEE Computer Society (2009)
20. Suzuki, D., Saeki, M.: Security evaluation of DPA countermeasures using dual-rail pre-charge logic style. In: Goubin, L., Matsui, M. (eds.) CHES 2006. LNCS, vol. 4249, pp. 255–269. Springer, Heidelberg (2006)
21. Tiri, K., Akmal, M., Verbauwhede, I.: A dynamic and differential CMOS logic with signal independent power consumption to withstand differential power analysis on smart cards. In: ESSCIRC 2002, pp. 403–406 (2002)
22. Tiri, K., Verbauwhede, I.: A Logic level design methodology for a secure DPA resistant ASIC or FPGA implementation. In: DATE 2004, pp. 246–251. IEEE Computer Society (2004)
23. Xilinx: Spartan-6 Libraries Guide for HDL Designs, October 2013
24. Yu, P., Schaumont, P.: Secure FPGA circuits using controlled placement and routing. In: CODES+ISSS 2007, pp. 45–50 (2007)

Side-Channel Protection by Randomizing Look-Up Tables on Reconfigurable Hardware

Pitfalls of Memory Primitives

Pascal Sasdrich[1]([✉]), Oliver Mischke[1,2], Amir Moradi[1], and Tim Güneysu[1]

[1] Horst Görtz Institute for IT-Security, Ruhr-Universität Bochum,
Bochum, Germany
{pascal.sasdrich,amir.moradi,tim.gueneysu}@rub.de
[2] Infineon Technologies AG, Chip Card and Security Division, Munich, Germany
oliver.mischke@infineon.com

Abstract. Block Memory Content Scrambling (BMS), presented at CHES 2011, enables an effective way of first-order side-channel protection for cryptographic primitives at the cost of a significant reconfiguration time for the mask update. In this work we analyze alternative ways to implement dynamic first-order masking of AES with randomized look-up tables that can reduce this mask update time. The memory primitives we consider in this work include three distributed RAM components (RAM32M, RAM64M, and RAM256X1S) and one BRAM primitive (RAMB8BWER). We provide a detailed study of the area and time overheads of each implementation technique with respect to the operation (encryption) as well as reconfiguration (mask update) phase. We further compare the achieved security of each technique to prevent first-order side-channel leakages. Our evaluation is based on one of the most general forms of leakage assessment methodology known as *non-specific t-test*. Practical SCA evaluations (using a Spartan-6 FPGA platform) demonstrate that solely the BRAM primitive but none of the distributed RAM elements can be used to realize an SCA-protected implementation.

1 Introduction

Side-channel analysis (SCA) exploits information leakage related to the device internals, for example by inspecting its power consumption [6]. Hence, the security provided by a cryptographic primitive can be easily overcome if the device is not equipped with any SCA countermeasures. Many different countermeasures against SCA attacks have been proposed in the past and can typically be classified as either masking or hiding [7]. With respect to signal-to-noise ratio hiding countermeasures aim at either increasing the noise by introducing noise generation resources [4,7] or reducing the signal by e.g., equalizing the power consumption [11]. The main concept behind masking is to randomize the processed values by adding random masks so that it should become impossible for an attacker to predict intermediate values. Despite many proposals, most fail to achieve the desired level of security due to the presence of glitches inside

S. Mangard and A.Y. Poschmann (Eds.): COSADE 2015, LNCS 9064, pp. 95–107, 2015.
DOI: 10.1007/978-3-319-21476-4_7

the combinatorial masked circuits (for example see [8,9]). Instead of masking combinatorial circuits, critical elements such as S-boxes can be realized as look-up tables that are dynamically randomized in memory. A realization of such an approach on FPGAs which randomizes the content of block RAMs (BRAM) has been presented in [4] and is known as Block Memory Content Scrambling (BMS).

Contribution: In this work we analyze the suitability of different Xilinx FPGA memory primitives to prevent first-order side-channel leakage by masked look-up tables. Besides using larger dual-port BRAM primitives (as used in the original BMS publication [4]), it is also possible to use smaller single-port BRAMs as well as distributed RAM elements which are realized in SLICE-M LUTs of modern Xilinx FPGAs [12]. With the introduction of Xilinx' Virtex-5 platform SLICE-M have become capable to hold 256 bits of memory that is a perfect fit for an 8×256-bit AES S-box. In particular RAM32M, RAM64M, and RAM256X1S are the primitives which can be used to build a randomly permuted (masked) S-box. Although reconfiguration time becomes notably shorter for smaller RAM module sizes, the total area requirements of each masked S-box increases.

For evaluation we apply the non-specific t-test as a general leakage assessment methodology [3] to analyze the SCA resistance of each scheme. We show that due to their intrinsic multi-LUT design, the distributed RAM elements still exhibit a first-order leakage so that they should not be used to implement masked designs. We conclude our work with presenting an efficient implementation of a small single-port BRAM-based design that achieves almost double the throughput of the original BMS scheme and still prevents first-order leakages.

Outline: This work is organized as follows: Sect. 2 introduces the underlying FPGA primitives, explains how they can be employed to realize randomized look-up tables and recalls the BMS scheme. In Sect. 3 our masked AES encryption designs are presented and their reconfiguration time, resource requirements, and throughput are compared. Practical evaluation of all implementation profiles is given in Sect. 4, before we conclude our work in Sect. 5.

2   Preliminaries

In this section we briefly describe memory primitives provided by Xilinx FPGAs and their application in order to build randomized look-up tables to protect cipher implementations against first-order DPA attacks. Afterwards, we restate the concept of Block Memory Content Scrambling (BMS) initially introduced in [4].

2.1   Memory Primitives

Modern Xilinx FPGAs provide several different memory primitives, e.g., distributed memory and general purpose block memory, that can be used to build randomly permuted look-up tables. Distributed memories are enabled only at special Slices (SLICE-M) by using the configuration registers within the Look-Up Tables (LUTs) as general purpose memory cells. Since this memory is usually constrained by the configuration size (between 16 and 64 bits), up to 4 LUTs of a single SLICE-M can be combined in order to build larger RAMs. For designs requiring even larger amounts of memory, FPGAs provide general purpose block memory (BRAM) with memory sizes between 8 Kbits and 32 Kbits. In the following we describe these memory primitives and their modes of operation in detail, focusing on their application as a randomized look-up table (see [12] for more information).

RAM32M. The RAM32M memory primitive is a multi-port random access memory with synchronous write but asynchronous read capability implemented in distributed memory using the configuration memory of all LUTs (and both outputs O6 and O5) of a single SLICE-M. It is organized as an 8-bit wide by 32 deep memory providing 4 individual read ports (each 2-bit wide) and a single write port (8-bit wide). If all read addresses are tied to the same value, this memory primitive becomes an 8×32 single port RAM.

RAM64M. In contrast to the RAM32M primitive, the RAM64M module is a multi-port random access memory with synchronous write and asynchronous read capability organized as 4-bit by 64 deep memory. This memory primitive also occupies 4 LUTs of a SLICE-M but only uses the outputs O6 of the LUTs. If all 6-bit wide address ports are tied to the same value, this memory becomes a 4×64 single port RAM.

RAM256X1S. Another option for distributed memory is RAM256X1S. This primitive is a single-port random access memory with synchronous write and asynchronous read capability placed in a single SLICE-M using all LUTs (combined by subsequent MUXF7 and MUXF8 multiplexer instances). A RAM256X1S provides an 8-bit wide address port and a 1-bit wide read and write port and is organized as a 1×256 single port RAM.

RAMB8BWER. The RAMB8BWER primitive is a true dual-port random access memory with synchronous read and write capability. Instead of using configuration memory of special LUTs as distributed memory, this RAM instance occupies a dedicated block memory primitive and offers 8 Kbits data storage in addition to a 1 Kbit parity memory. It is possible to define different options and widths for the read and write ports changing the memory configuration from 1×8 Kbits up to 9×1 Kbit. The embedded input register causes this primitive to always require a clock cycle to read from an address (synchronous). In addition, the output port can use an additional embedded register in order to buffer the memory output leading to two clock cycles latency for a read operation.

2.2   Randomized Look-Up Tables

Many symmetric ciphers use S-boxes, often represented by simple look-up tables, in order to include non-linearity into the encryption scheme. In FPGAs, this S-boxes can efficiently be realized either using LUTs (as well as distributed memories) or block memories depending on their size as well as the available resources.

SCA attacks target an intermediate value of a cipher, e.g., a part of the non-linear layer. The predicted intermediate values, usually the input or output of a known S-box, in addition to a hypothetical power model contribute in a statistical analysis of e.g., power consumption traces in order to reveal the associated secret. In order to avoid side-channel leakages, hardware designers need to apply dedicated countermeasures e.g., masking. These countermeasures aim at randomizing intermediate values of a cipher implementation using uniformly-distributed random data (masks). In particular, the non-linear layer in terms of look-up tables such as S-boxes (or T-Tables) has to be adapted depending on the taken random mask. Usually this is done by scrambling the S-box content based on an input mask m and adding an output mask n to the content (Boolean masking), so that the masked S-box S' is precomputed as:

$$S'(x \oplus m) = S(x) \oplus n$$

As mentioned before, look-up table based S-boxes can be implemented using distributed or block memories. Due to their reconfiguration feature, the above-presented memory primitives can be employed to implement randomized look-up tables as well. Figure 1 exemplarily shows a part of the structure of an AES S-box using RAM32M (Fig. 1a), RAM64M (Fig. 1b), RAM256X1S (Fig. 1c) and RAMB8BWER (Fig. 1d) memory primitives.

Each of the distributed memory designs presented in Fig. 1 realizes one bit of the AES S-box. Each of them receives an 8-bit input S_{in}, and provides one output bit S_{out}. Depending on their read/write port width the configuration to update the look-up table is defined. For example, the content of 8 bits of a RAM32M can be updated in one clock cycle (Fig. 1a) while at most 4 bits of RAM64M and 1 bit of RAM256X1S can be simultaneously updated. This clearly affects the efficiency of the update (reconfiguration) process. Respectively, extra components, i.e., the multiplexers in Fig. 1, have to be placed out of the SLICE-M to build a 1×256 memory. With respect to this issue RAM256X1S is the most efficient one while the time required to update its content is considerably higher than the other distributed memory primitives.

2.3 Block Memory Content Scrambling

The main idea of BMS is to store two S-/T-Tables in parallel into a dual-port block memory where one is called active context and the other one passive. While the active context is used for the encryption process via one port of the BRAM, the passive context is scrambled by means of the other port. During the scrambling process the already masked data is read from the active context, and updated by a given fresh mask before it is written to the passive context. After the encryption and the memory content scrambling process finished, the contexts are swapped i.e., the passive context becomes active and is used for the encryption process while the active context becomes passive and is updated

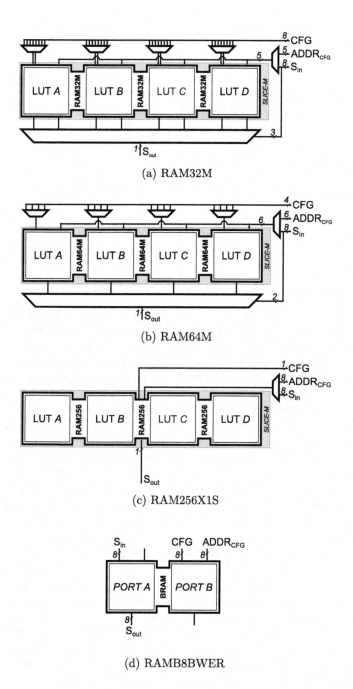

(a) RAM32M

(b) RAM64M

(c) RAM256X1S

(d) RAMB8BWER

Fig. 1. Randomized look-up tables using different memory primitives

using a new (random) mask. This scrambling scheme exploits the true dual-port capability of BRAM in order to randomize look-up tables such as S-boxes or T-Tables without affecting the throughput of the encryption scheme. Despite many advantages, this scheme comes with

- Area overhead, since it doubles the memory requirements because every look-up table has to be stored twice (active and passive), and
- Additional latency for a mask update process, as the scrambling (updating) process needs 512 clock cycles. Hence it often happens that the consecutive encryptions share the masks since the scrambling process is not finished when the second plaintext is given.

3 Design

This section briefly explains the underlying masking scheme of our AES implementation and its basic hardware architecture. Afterwards, different approaches using the distributed memory and the block memory primitives are compared.

3.1 Masking Architecture

The architecture of our design of the AES-128 encryption function (for a Spartan-6 FPGA) is shown in Fig 2. We opted to implement an incremental and round-based architecture and derive the round keys on the fly. The data path has a width of 128 bits, and the SubBytes layer consists of 16 parallel reconfigurable S-boxes. ShiftRows and MixColumns (in parallel on all 4 columns) are applied jointly at one clock cycle.

In contrast to the originally proposed BMS scheme, our design follows an approach based on an update-prior-to-encryption fashion. Thus, before each encryption the randomized look-up tables are regenerated. During each encryption the masks stay constant. In other words, the same masks are used for all cipher rounds during one encryption. The initial plaintext is masked with $(m \oplus m')$ while all round keys are masked with m' (m and m' independent of each other and each 128-bit). Therefore, after the key addition the SubBytes input mask is m (see Fig. 2). The randomized look-up tables (masked SubBytes) are configured with m as the input mask and $SR^{-1}(MC^{-1}(m \oplus m'))$ as the output mask. Applying the ShiftRows and MixColumns operations transforms the mask again to $(m + m')$ as the mask of the round output. Hence, after each cipher round the input to the next round is masked with $(m \oplus m')$ and no mask correction (see [2,10]) is required. For the last round, the MixColumns operation is omitted and the returned ciphertext is masked with $MC^{-1}(m \oplus m') \oplus m'$.

Reusing the masks for all cipher rounds has a known drawback if the round register consecutively stores the intermediate values with the same mask. In such a case, the leakage associated to the register update, e.g., a Hamming distance (HD) model, is easily extractable. If $x \oplus m$ and $y \oplus m$ are consecutively stored in a register,

$$HD(x \oplus m, y \oplus m) = HW(x \oplus y)$$

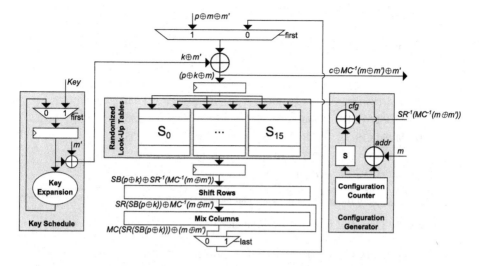

Fig. 2. Round-based AES implementation with randomized look-up tables

is independent of the mask. Hence, we avoid such an issue by surrounding each S-box with two register stages, one before and one after the SubBytes operation (see Fig. 2). At power-up both registers are precharged with 0, and at only one clock cycle the input multiplexer passes the masked plaintext ($p \oplus m \oplus m'$). Since one of the register stages therefore holds some value depending on a random mask of a previous encryption, the correct encryption rounds are interleaved with random (dummy) operations.

Employing this technique leads to reduced throughput due to the prior look-up table update phase as well as the fact that each cipher round requires two clock cycles. However, compared to BMS [4] our design reduces the area overhead as well as the amount of required randomness to 256-bit per encryption (m and m'). Further, this scheme is suitable for the distributed memory primitives as well as for the block memory which allows a fair comparison. In case the block memory is used, the registers (before and after the SubBytes) are removed. Instead, the input and output registers of the block memory are employed as the two-stage state registers.

3.2 Comparison of S-box Designs

Table 1 provides a comparison of area and time requirements of the randomized look-up tables using different memory primitives and the associated configuration logic and in Table 2 we give an overview of the resource requirements of the entire AES encryption as well as an estimation of the maximum frequency and throughput. Compared to the originally proposed BMS scheme, our masked design based on the block memory (RAMB8BWER) halves the reconfiguration time, hence nearly doubling the maximum throughput. In case the distributed memory primitives are employed, the maximum frequency can even be increased except for the RAM32M

Table 1. Comparison of S-boxes for different memory primitives

Memory primitive	Subbytes			Configuration	
	Logic	Dist. mem.	Block mem.	Logic	Memory
	(LUT)	(LUT)	(BRAM16)	(LUT)	(FF)
BRAM (BMS)	none	none	16	1706*	1169*
RAMB8BWER	none	none	8	298	8
RAM256X1S	128	512	none	298	8
RAM64M	768	512	none	727	6
RAM32M	1920	512	none	1222	5

* These values are based on a Virtex-II Pro implementation and taken from [4]. For a Spartan-6 the resulting design would be slightly smaller.

Table 2. Time and resource requirements of entire AES (encryption only)

Memory primitive	AES (Encryption only)			Reconfig.	Maximum	Maximum
	Logic	Memory		Time	Frequency	Throughput
	(LUT)	(FF)	(BRAM)	(Cycles)	(MHz)	(MBit/s)
BRAM (BMS)	2888	2351	16	512*	147.0	35.4
RAMB8BWER	1284	415	8	256	148.0	68.6
RAM256X1S	1796	543	0	256	166.1	77.0
RAM64M	2849	541	0	64	162.3	247.3
RAM32M	4512	540	0	32	147.6	363.3

* Reconfiguration can be done in parallel when reusing the mask for multiple encryptions without affecting the throughput. For a fair comparison we avoid the mask reuse in BMS as well.

due to its more complex reconfiguration circuit. Besides, the RAM32M leads to the highest throughput as its reconfiguration time is extremely shorter than the others. Note that in the reported performance figures we omitted the area required for the generation of the random masks.

4   Evaluation

We employed a SAKURA-G platform [5], i.e., a Spartan-6 FPGA, for practical side-channel evaluations. The power consumption traces have been measured by means of a LeCroy WaveRunner HRO 66Zi oscilloscope with a $1\,\Omega$ resistor in the V_{dd} path capturing the embedded amplifier output of the SAKURA-G. We recorded the traces at a sampling rate of $1\,\mathrm{GS/s}$ and the bandwidth limit of $20\,\mathrm{MHz}$ while the design was running at a low clock frequency of $3\,\mathrm{MHz}$ to reduce the noise caused by the overlap of the power traces.

4.1   Non-Specific Statistical t-test

In order to examine the resistance of our designs we applied the leakage assessment methodology (t-test) of [3]. The most general form of such a test – known as *non-specific* t-test – investigates the existence of a first-order leakage independent of any power model as well as any intermediate value. In such a test a certain plaintext is selected, and during the measurements the chosen plaintext or a random one is given to the encryption module in a randomly-interleaved fashion. For all the measurements the key is kept constant. Therefore, this test is also called *fix vs. random* t-test. As the next step the traces are categorized into two groups G_1 and G_2 based on their associated (fix or random) plaintext. By comparing the means of these groups, we can examine the dependency of the traces (leakages) to the processed values related to the given plaintexts. Such a comparison can be fairly performed by means of a Welch's (two-tailed) t-test as

$$ t = \frac{\mu(T \in G_1) - \mu(T \in G_2)}{\sqrt{\frac{\delta^2(T \in G_1)}{|G_1|} + \frac{\delta^2(T \in G_2)}{|G_2|}}}, $$

where μ and δ^2 denote the sample mean and the sample variance respectively, and $|.|$ the cardinality.

As the final step the obtained t with the corresponding *degree of freedom*[1] is given to the cumulative Student's t distribution function to achieve a quantitative value as the probability of the *null hypothesis* being valid. Such a hypothesis is the assumption that the samples in the groups G_1 and G_2 were drawn from the same population, i.e., the two groups are not distinguishable. However, for simplicity a threshold for the t-test result as $|t| > 4.5$ is usually selected to reject the null hypothesis and conclude that the means of the groups are distinguishable, hence there exists a leakage.

It is noteworthy that the scenario explained above should be repeated at each sample point of the power traces independently, hence a first-order univariate evaluation. On one hand, when the result of a test is positive, the value of the t statistic gives the level of confidence that there exist a first-order leakage, but it does not provide any information about the difficulty or easiness of an attack exploiting such a leakage. On the other hand, in case a non-specific t-test reports no leakage, such a conclusion is only correct with respect to the selected fix plaintext as well as the number of used traces. Changing the fix plaintext and increasing the number of traces can change the result of the test. The same evaluation scheme has also been applied in [1].

4.2   Results

In the following we present the results of the security evaluation concerning side-channel resistance of randomized look-up tables using the introduced memory primitives by applying the above-explained non-specific t-test. Since we identified four potential memory elements (see Sect. 2.1), the evaluations are grouped into four different profiles respectively.

A sample trace of the profile built from RAM64M modules is shown in Fig. 3. Note that all our measurements cover only the time period related to the encryp-

---

[1] see [1] and [3].

tion, and we ignored to measure the power consumption when the reconfiguration of the look-up tables is in process (prior to each encryption). As explained
in Sect. 3, we kept the design architecture of all profiles the same. Hence the
power traces of other profiles look like the same, but for the design profile with
RAM32M the traces show slightly higher peak-to-peak amplitude due to its more
complex architecture regarding the extra multiplexers out of the RAM slices.

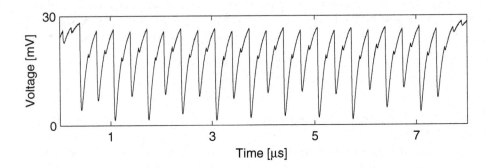

Fig. 3. Sample trace

For each profile we collected at least 1 million traces for a non-specific t-test.
During all the measurements fresh masks are randomly generated by means of
an AES engine running in counter mode prior to each encryption, i.e., no mask
is reused. The masked plaintext in addition to the corresponding masks are sent
from the control FPGA to the target FPGA (SAKURA-G). After finishing the
look-up table reconfiguration followed by the encryption process on the target
FPGA, the masked ciphertext is sent back to the control FPGA, where it is
unmasked for a consistency check.

Profile A: Tiny RAM (RAM32M). By means of this profile we evaluate the
leakage of the randomized look-up table realized by RAM32M memory primitives. Although this variant has the highest resource consumption, it offers the
best throughput. Figure 4a shows the result of the corresponding non-specific
t-test using 1 million traces (i.e., about 500 000 traces of encrypting the fix
plaintext and the rest for the random ones). Unexpectedly the test exhibits
first-order leakages. Indeed, the t statistics are much higher than the threshold,
that confidently argue the vulnerability of the design.

Profile B: Small RAM (RAM64M). The result of the same test on the
second profile, i.e., the one where the randomized look-up tables are implemented
by RAM64M instances, is shown in Fig. 4b. We observed the same issue, i.e.,
unexpected first-order leakages. Interestingly, the amount of leakage is higher
compared to that of Profile A, although its S-box design is more compact.

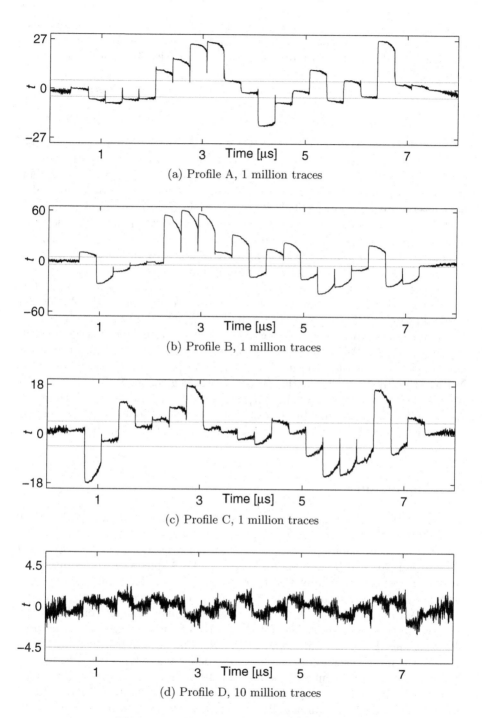

(a) Profile A, 1 million traces

(b) Profile B, 1 million traces

(c) Profile C, 1 million traces

(d) Profile D, 10 million traces

Fig. 4. First-order non-specific t-test results

Profile C: Large RAM (RAM256X1S). The most compact and dense implementation for a randomized look-up table using Distributed Memory (i.e., the RAM256X1S memory primitives) on a Spartan-6 FPGA, places the a complete single AES S-box and the subsequent registers into only 8 slices. However, the same as the two former design profiles a first-order leakage is still detectable which can be seen in Fig. 4c.

Profile D: Block RAM (RAMB8BWER). As the last profile we evaluated the application of block RAMs instead of the Distributed Memory. Since each block RAM internally has a register stage for the input and an optional one for the output, by employing a RAMB8BWER instance for each S-box we used also both internal registers of the block RAM and avoided the external registers used in the other profiles (see Fig. 2). Since we did not observe any first-order leakage using the same number of traces as used for the other profiles, we performed the evaluation using 10 million traces. The corresponding result is shown in Fig. 4d indicating the ability of the design to prevent any first-order leakage.

In fact, the results we presented above infer the pitfall of using distributed memories (of FPGAs) to realize randomized (masked) look-up tables. While the internal architecture of such memory primitives is not completely clear to us, we are confident that the observed leakage is due to the internal multiplexers of such memory modules. We should highlight that the randomized look-up tables (in our designs) receive only the masked inputs and provide the masked outputs. Neither the input mask nor the output mask is given to the memory module. Further, the input masks and output masks are independent of each other. As a result – also confirmed by the evaluation result of Profile D – the exhibited leakage is purely related to the internal architecture of the distributed memory modules.

5   Conclusion

In this work we have given a comparative study on the suitability of Xilinx FPGA memory primitives to implement a side-channel countermeasure based on randomized (masked) look-up tables. We have shown that the use of distributed RAM primitives like RAM32M, RAM64M, and RAM256X1S causes an otherwise secure scheme to exhibit first-order side-channel leakage. Such unexpected leakage is due the internal architecture of the distributed memory primitives (SLICE-M). Since except [12] there is no other public document on the details of such modules, we cannot localize the source of such leakage. When keeping the very same design but only replacing the distributed RAMs by small BRAMs to store the masked tables, no leakages were detected applying the general non-specific leakage assessment methodology on 10 million captured power traces.

Our design solution using block memory (RAMB8BWER) achieves almost double the throughput compared to the original BMS mainly because of the reduced reconfiguration time of the masked S-boxes. It also requires less randomness. The BMS scheme is a T-table implementation which requires 16×32

random bits to mask the T-tables output while we only require 2×128 bits of randomness. The reason for this difference is that we are only implementing the 8×8 AES S-box as masked tables (compared to 8×32 T-tables) while the other parts (all linear) of the encryption are implemented by combinatorial logic.

References

1. Bilgin, B., Gierlichs, B., Nikova, S., Nikov, V., Rijmen, V.: Higher-order threshold implementations. In: Sarkar, P., Iwata, T. (eds.) ASIACRYPT 2014, Part II. LNCS, vol. 8874, pp. 326–343. Springer, Heidelberg (2014)
2. Bringer, J., Chabanne, H., Le, T.: Protecting AES against side-channel analysis using wire-tap codes. J. Crypt. Eng. **2**(2), 129–141 (2012)
3. Goodwill, G., Jun, B., Jaffe, J., Rohatgi, P.: A testing methodology for side-channel resistance validation. In: NIST Non-Invasive Attack Testing Workshop, Nara (2011)
4. Güneysu, T., Moradi, A.: Generic side-channel countermeasures for reconfigurable devices. In: Preneel, B., Takagi, T. (eds.) CHES 2011. LNCS, vol. 6917, pp. 33–48. Springer, Heidelberg (2011)
5. Guntur, H., Ishii, J., Satoh, A.: Side-channel attack user reference architecture SAKURA-G. In: GCCE 2014. IEEE Computer Society (2014). http://satoh.cs.uec.ac.jp/SAKURA/index.html
6. Kocher, P.C., Jaffe, J., Jun, B.: Differential power analysis. In: Wiener, M. (ed.) CRYPTO 1999. LNCS, vol. 1666, pp. 388–397. Springer, Heidelberg (1999)
7. Mangard, S., Oswald, E., Popp, T.: Power Analysis Attacks - Revealing the Secrets of Smart Cards. Springer, New York (2007)
8. Mangard, S., Pramstaller, N., Oswald, E.: Successfully attacking masked AES hardware implementations. In: Rao, J.R., Sunar, B. (eds.) CHES 2005. LNCS, vol. 3659, pp. 157–171. Springer, Heidelberg (2005)
9. Moradi, A., Mischke, O., Eisenbarth, T.: Correlation-enhanced power analysis collision attack. In: Mangard, S., Standaert, F.-X. (eds.) CHES 2010. LNCS, vol. 6225, pp. 125–139. Springer, Heidelberg (2010)
10. Nassar, M., Souissi, Y., Guilley, S., Danger, J.: RSM: a small and fast countermeasure for AES, secure against 1st and 2nd-order zero-offset SCAs. In: DATE 2012, pp. 1173–1178. IEEE (2012)
11. Tiri, K., Verbauwhede, I.: A logic level design methodology for a secure DPA resistant ASIC or FPGA implementation. In: 2004 Design, Automation and Test in Europe Conference and Exposition (DATE 2004), pp. 246–251. IEEE Computer Society, 16–20 February 2004, Paris, France (2004)
12. Xilinx. Spartan-6 Libraries Guide for HDL Designs (UG615 v 14.1), April 2012. http://www.xilinx.com/support/documentation/swmanuals/xilinx141/spartan6hdl.pdf

Timing Attacks and Countermeasures

A Faster and More Realistic *Flush+Reload* Attack on AES

Berk Gülmezoğlu[(✉)], Mehmet Sinan İnci, Gorka Irazoqui,
Thomas Eisenbarth, and Berk Sunar

Worcester Polytechnic Institute, Worcester, MA, USA
{bgulmezoglu,msinci,girazoki,teisenbarth,sunar}@wpi.edu

Abstract. Cloud's unrivaled cost effectiveness and on the fly operation versatility is attractive to enterprise and personal users. However, the cloud inherits a dangerous behavior from virtualization systems that poses a serious security risk: resource sharing. This work exploits a shared resource optimization technique called memory deduplication to mount a powerful known-ciphertext only cache side-channel attack on a popular OpenSSL implementation of AES. In contrast to the other cross-VM cache attacks, our attack does not require synchronization with the target server and is *fully asynchronous*, working in a *more realistic* scenario with much weaker assumption. Also, our attack succeeds *in just 15 seconds* working across cores in the cross-VM setting. Our results show that there is strong information leakage through cache in virtualized systems and the memory deduplication should be approached with caution.

Keywords: Asynchronouos cross-VM attack · Memory deduplication · Flush and reload · Known ciphertext attack · Cache attacks

1 Introduction

Cloud computing and virtualization is popular more than ever with large companies like Microsoft, Google, Amazon, IBM, Oracle, Rackspace and many others investing billions of dollars trying to get a foothold in this new area of lucrative business. This rapid increase in the number of cloud service providers is directly related to the emergence of server-less companies like Netflix, Dropbox, Instagram, Pinterest, Reddit, Imgur and many others that are using commercial cloud infrastructure [10]. Instead of buying expensive servers without knowing exactly how many of them they need, and then hiring IT personnel to maintain those servers, these fast growing companies have chosen to use public cloud systems to maintain their software and services.

The opportunities of using the commercial cloud are fairly obvious however, threats are not. Sharing a physical system between users reduces the cost while increasing the utilization hence the productivity. The isolation between the Virtual Machines (VM) in these systems is maintained by the Virtual Machine Manager (VMM) at the software level. However, software layer confinement

© Springer International Publishing Switzerland 2015
S. Mangard and A.Y. Poschmann (Eds.): COSADE 2015, LNCS 9064, pp. 111–126, 2015.
DOI: 10.1007/978-3-319-21476-4_8

techniques that force the *sandboxing* does not guarantee complete isolation and cannot ensure the prevention of data leakage from one VM to the other. The most common source of information leakage across VM boundaries is the shared cache and the memory of the underlying physical system. Particularly memory deduplication allowed researchers to mount attacks that threaten both the user privacy and the security of the cryptographic systems.

In 2009, Ristenpart et al. [24] showed that it is possible to co-locate with a target on a cloud environment, namely Amazon EC2, and extract keystrokes from the co-located VM. In 2011, by exploiting the Kernel Samepage Merging (KSM), Suzaki et al. [25] was able to detect processes like sshd, apache2, IE6 and Firefox running on a co-resident VM. The significance of this study is that it is possible to not just detect the existence of a target VM, but also detect running processes.

Recently in 2013 Yarom et al. [29] applied the *Flush+Reload* attack across VMware VMs to recover a RSA key. Later in 2014, Irazoqui et al. [14] used Bernstein's AES cache timing attack to partially recover an AES key from various AES crypto library implementations in a cross-VM setting under XEN and VMware ESXI hypervisors. Also in 2014, Irazoqui et al. [15] implemented a cross-VM access driven cache attack on AES in a VMware ESXI system using the *Flush+Reload* attack.

Our Contribution. In this work, we implement for the first time a known-ciphertext cross-VM attack on AES using the *Flush+Reload* method and use three distinct data analysis methods to fully recover the secret key with varying encryption observations for different scenarios. For the attack, we take advantage of VMware ESXI' s memory deduplication mechanism called the Transparent Page Sharing. The attack is mounted on a multi-core high-end server, a specification found commonly on commercial cloud systems and does not require the attacker and the victim to be running on the same physical CPU core. Compared to the attack in [12], our attack is minimally invasive and works with less assumptions since the attacker does not need to control or exploit in any way the target process execution. Also compared to [15], the new attack does not assume to have access to the encryption server and works only by listening to the encryption server via cache covert channel and obtaining the ciphertexts from the network channel.

In summary, this work

- For the first time, mounts a cross-VM, **known-ciphertext only** AES key recovery attack using the *Flush+Reload* technique
- Improves upon the previous cross-VM AES cache attacks by flushing **in between** the encryption rounds
- Presents three distinct analysis methods that can be adapted to any table-based block ciphers

2   Cache Side-Channel Attacks

Cache side-channel attacks exploit microarchitectural leakages stemming from memory access time variations, e.g. when the data is retrieved from small faster memories called caches as opposed to slow bulk memories. Caches are useful due to two main principles, i.e. the *temporal* and *spatial locality*. *Temporal locality* predicts that recently accessed memory locations are likely to be accessed soon again, while *spatial locality* predicts that data located nearby the accessed memory locations are also likely to be accessed soon. In general, caches hold not only recently accessed data, but also an entire cache line containing data in nearby locations. In modern CPUs, caches are organized into multiple levels L1, L2 and L3 where the first two levels are smaller and core exclusive, while the last level is considerably larger and is shared across cores. While retrieving data from any level of the cache is faster than retrieving the same data from the main memory, higher levels of the cache are even faster than lower levels. L1 being the fastest, L3 the slowest, different cache levels have different access times which enable attacks like *Prime+Probe* to distinguish between accesses to the L1 cache and to the L3 cache. In addition, the last level of cache is shared between all cores, giving the attackers the opportunity to use it as a covert channel between cores and mount attacks such as *Flush+Reload*.

2.1   Related Work

The first theoretical consideration on the extraction of information via cache memories was demonstrated by Hu [13], whereas 6 years later Kelsey et al. [18] expanded this consideration by suggesting the presence of cache leakage due to the hit/miss ratio. Following up Kelsey's work, Page described a theoretical chosen plaintext attack based on the collection of cache profiles [23]. One year later Tsunoo et al. [26] proposed the first practical implementation of cache attacks on the DES cryptographic algorithm.

The first practical cache side-channel attack on AES appeared in 2005 by Bernstein [7] showing that the table look up operations from different cache lines have different access times in an AES encryption. Further, he showed that using this cache access time information, an adversary can recover secret encryption key from a popular AES implementation, i.e. the OpenSSL cryptographic library. In a similar attack, Osvik et al. [22] presented two spy processes that are able to monitor the cache usage: evict+prime and prime+probe. Although the latter one proved to be significantly more efficient, both spy processes recovered the AES encryption key used by an OpenSSL server. A few months later, Bonneau and Mironov [8] and Acıiçmez and Koç [5] presented new attacks targeting AES that exploited internal table look up collisions in the cache during the last and first rounds respectively. In spite of the prominent successful attacks on symmetric key cryptography, public key cryptography was also considered a popular target for cache side-channel attacks. Indeed in 2007, Acıiçmez demonstrated the usage of the prime and probe spy process in the instruction cache against a RSA encryption.

Cloud computing systems became the next challenge for side-channel attack researchers. In 2009, after Ristenpart et al. [24] demonstrated that they were able to co-locate an attacker's virtual machine (VM) with a potential victim's VM with a success probability of 40 % in the Amazon EC2 cloud. Even further, the authors managed to recover keystrokes from the co-resident victim's VM, showing that the cache side-channel attacks are both practical and applicable to real world scenarios. The possibility of co-location fueled further research on new cache side-channel techniques and cache leakages in VMs. For instance, in 2011 Chen et al. improved over the previous RSA attacks in the instruction cache [9] while Gullasch et al. discovered a new side-channel technique that would later be called the *Flush+Reload* [12]. The *Flush+Reload* attack recovered AES secret keys by taking control of the Completely Fair Scheduler (CFS) [3,16]. At the same time, previous side-channel techniques such as *Prime+Probe* were also adapted to work in virtualized settings by Zhang et al. [30,31]. They utilized a spy process to detect co-resident tenants and to recover El Gamal encryptions keys. More recently, Yarom and Falkner [29] applied the *Flush+Reload* technique to recover, for the first time, RSA encryption keys across VMware and KVM VMs. Shortly later Benger et al. [6] demonstrated the viability of the *Flush+Reload* technique to recover ECDSA encryption keys. Finally, Irazoqui et al. [14,15] recovered AES keys in virtualized environments with Bernstein's attack and the *Flush+Reload* technique.

2.2 Memory Deduplication

Memory deduplication is an OS memory optimization technique that allows the OS to keep only a single copy of a data in the memory when multiple processes are using the same data. While this feature is useful in native execution, it is even more useful in virtualized setting where many VMs use the same OS and/or the same software.

Hence, to reap the benefits of the deduplication, VMMs have also implemented memory deduplication techniques to allow more VMs to run on the same physical machine. For this, the VMM recognizes identical and redundant memory copies by first checking their hash values and then performs a bit-by-bit comparison. If the memory content is determined to be shared by more than one process/VM the memory manager removes multiple copies from the memory. Note that even though this deduplication process is only performed on shareable memory pages like shared libraries, shared libraries are used in many software packages. Memory deduplication methods are especially effective when hosting multiple processes, as is the case in virtualized systems. Consequently, VMMs like VMware [27,28] and KVM [4,17] implement variations of memory deduplication, i.e. Transparent Page Sharing (TPS) and Kernel Samepage Merging (KSM), respectively. While the memory saving optimization techniques improve the performance they also create a covert channel that a malicious VM can exploit. In fact, memory disclosure attacks [25] and side-channel attacks [6,15,29] have been proposed taking advantage of memory deduplication techniques in the cloud.

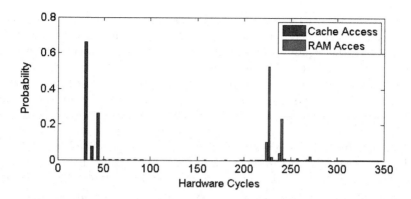

Fig. 1. Data access time in hardware cycles when the data is located in the cache and in the memory

2.3   The *Flush+Reload* Side-Channel Attack

The *Flush+Reload* is a trace driven cache side-channel attack that was first used in [12], but acquired its name in [29]. The attack is based on shared memory leakage coming from deduplication processes. One of the main advantages of the *Flush+Reload* spy process is that it does not require the attacker to be core co-resident with the victim and works in a cross-core scenario *as long as a shared last level cache exists*. The attack is carried out in 3 main stages:

Flush Stage: In this stage the attacker flushes one or more of the desired memory locations from the cache using the `clflush` command. More precisely, `clflush` evicts the desired memory locations from the entire cache hierarchy, i.e. even from the non-shared cache hierarchies if the last shared level cache is inclusive. Indeed this is the main reason why the attack is applicable across cores.

Victim Access Stage: In this stage, the attacker waits for sufficient time for the victim to use (or not use) the memory locations that he has flushed in the previous stage.

Reloading Stage: In the final stage, the attacker reloads the previously flushed memory locations, measuring the reload time for each one of them. If the victim accessed one of the flushed memory lines, due to the inclusiveness of the shared level cache, they will not only be loaded in the upper level caches but also in the shared level cache. Thus, the attacker will measure a lower reload time compared to data accesses to the main memory since the line will be retrieved from the cache. However if the victim did not access to the data flushed in the first stage, the data will still reside in the memory, causing a higher reload time in this reload stage. The different distributions for a memory block accessed from the

L1 cache and a memory block accessed from the main memory can be observed in Fig. 1. It can be concluded that *Flush+Reload* offers a high distinguishable covert channel due to the significantly different distributions. However, the execution of microarchitectural side-channel techniques can suffer from multiple sources of noise that can be observed in two different ways. The first is a measurement inaccuracy: noise can be introduced by the microarchitecture, by the OS and by the VMM. Often, this noise results in a moderate increase in the number of cycles. However, if e.g. a context switch happens during start and end of a measurement, the value might be off several orders of magnitude. This can be handled by introducing a threshold. Note that such outliers have a significant impact on higher order statistical moments if not filtered out. This said, even with a reasonable threshold, the noise is definitely not Gaussian, possible better described by ExGaussian distributions. The second effect of noise is independent of the measurement process. This happens if a cache line is loaded or evicted by another process. In this case, the source of the timing changes, in addition to the noise introduced during measurement.

3   Attack Description

Our attack uses the side-channel technique known as *Flush+Reload* to monitor accesses to memory blocks. The *Flush+Reload* is applicable in the cross-VM setting if deduplication is enabled by the hypervisor and the monitored part of the memory is deduplicated. The latter is true if the monitored data is marked as shared (as is the common case for all crypto libraries) and the hypervisor has detected the duplicated data referenced from within both VMs. Also, different than the attack in [15], we utilize a separate AES detection step to detect the AES execution on the co-located target VM and eliminate the synchronization requirement with the server through the plaintext generation. This makes the proposed attack much more practical. We access the AES function memory address to detect the beginning of AES execution by *Flush+Reload* method. The reason why we access the memory location instead of simply running AES is that accessing a single memory location is much faster than running AES, allowing a higher attack resolution.

3.1   A Single Cache Line Attack on AES

The adversary monitors accesses to a single block of one of the T tables used in the last round of AES. In addition to the information t whether the T Table was accessed, the adversary needs to know the corresponding ciphertext c (or plaintext for a first-round attack). That is, we assume the adversary is able to collect several tuples $\langle c, t \rangle$. The monitored memory block corresponds to n T table entries $\mathbb{T}$ known to the adversary. For a monitored ciphertext byte C_i, these entries correspond to n T table outputs S_i, which are mapped one-to-one to n ciphertext byte values through addition with the key. Hence, $c_{i,j} = k_i \oplus s_{i,j}$,

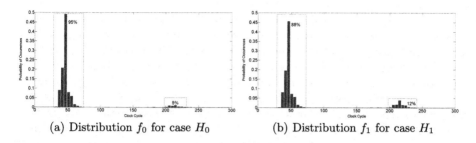

(a) Distribution f_0 for case H_0 (b) Distribution f_1 for case H_1

Fig. 2. Leakage distributions f_0 and f_1 if hypotheses H_0 and H_1 are correct. The measurements were taken in an intel i5 2430M CPU in SSA scenario.

where i is a byte position (ignoring the shift rows operation) and j indicates different values. If $s_{i,j}$ is equal to one of the values of the monitored T table memory block, i.e. $s_{i,j} \in \mathbb{T}$, then the monitored memory block will be accessed hence loaded to the cache. We will refer to this case as H_0. However, if $s_{i,j} \notin \mathbb{T}$, i.e. $s_{i,j}$ takes a value stored in a different memory block, then the monitored memory block is not loaded. Nevertheless, since each T table is accessed l times, there is still a high probability that the memory block was loaded by any of the other accesses. In fact, the probability that a memory block is not accessed during an encryption is given as: $\Pr[\text{no access to } T_j] = (1 - n/256)^l$. We will refer to this event as H_1.

For AES-128 in `OpenSSL 1.0.1g`, $n = 16$ and $l = 40$ per T_j, and therefore $100\% - \epsilon_0$ of reloads are expected to come from the cache in H_0, and only $92\% + \epsilon_1$ for H_1, where ϵ_i are noise terms. Hence, a side-channel containing information about memory/cache accesses will feature differing leakage distributions f_0 and f_1 for cases H_0 and H_1, respectively. To distinguish H_0 from H_1 the *Flush+Reload* method can be applied. In fact, using the *Flush+Reload* method, one can, with high probability, distinguish a cache access from a memory access as seen in Fig. 1. In our scenario (as described in Sect. 4) the leakage distributions f_0 and f_1 are depicted in Fig. 2. The distributions are derived from the reload times measured by the *Flush+Reload* attack. The first peak in both distributions (at around 35 cycles) corresponds to a noisy cache reload, and the second peak (at around 220 cycles) corresponds to a memory reload. Since f_0 corresponds to H_0 and hence has more cache reloads than f_1, these distributions are distinguishable. This leakage was successfully exploited in [15].

3.2 Distinguishers for the AES Attack

To process the side-channel data, we describe and compare three distinguishers. The distinguishers we present here analyze one byte of the ciphertext c together with the access time t to the corresponding T table block to recover one byte k of the last round key.

As described earlier, our observations are split into two sets according to a hypothesis. If this hypothesis is correct, the resulting leakage distributions

f_0 and f_1 for the two sets differ and hence—with sufficiently many observations—become distinguishable. For wrong key guesses, however, the hypotheses will be invalid, and both sets will sample from the same mixed distribution, making them indistinguishable. To detect whether samples for hypotheses H_0 and H_1 are actually from different distributions, we can apply several distinguishers. In the following we propose three distinguishers. The probably most common distinguisher is based on the difference of the means of the two distributions [11, 20]. As for the zero-value DPA [21], our hypothesis deviates from a single-bit prediction, yet, the test still just distinguishes two cases. Similarly, the variance test uses a statistical moment to distinguish the two distributions [11, 19, 20]. The last distinguisher applies a *miss counter*, as in [15]. The list is neither exhaustive, nor do we make an optimality claim. The latter is interesting future work that needs to be preceded by a better understanding and analysis of the underlying noise characterization, as noise can come from several different sources and is far from being Gaussian.

For the following descriptions we refer to the average miss counter value for H_i as $\overline{ctr}_{H_i}$, whereas we refer to the difference of means and difference of variances for f_i as $\overline{\tau}_{H_i}$ and var τ_{H_i}, respectively.

Miss-counter Based Distinguisher. This distinguisher counts and compares the memory block misses for the two cases H_0 and H_1. Ideally, there should be no misses for H_0, as the memory block must have been accessed by the AES execution. To establish a *miss counter*, reload timings are converted to either a hit (0) or a miss (1), depending on whether the value is above or below a threshold access time. As seen in Fig. 1, a good threshold for our processor and probing code is 130 cycles. Since H_1 contains significantly more values than H_0, we compare the relative counters instead of absolute ones. Our distinguisher becomes:

$$\mathcal{D}_{miss\_ctr} = \arg\max_{\hat{k}}\left(\overline{ctr}_{H_1} - \overline{ctr}_{H_0}\right)$$

Difference of Means Distinguisher. The difference of means distinguisher approximates the means of the two distributions and outputs their difference in cycles.

$$\mathcal{D}_{means} = \arg\max_{\hat{k}}(\overline{\tau}_{H_1} - \overline{\tau}_{H_0})$$

Since H_0 should feature more cache accesses than H_1, $\overline{\tau}_{H_0}$ is expected to be smaller, i.e. the biggest positive difference corresponds to the most likely key hypothesis. Welch's t-test distinguisher (which divides the means with their respective variance) can be equally well applied to guess the correct key. Indeed, Welch's t-test is commonly applied to check two hypothesis where two gaussian distributions have different means and variances. In this work, we studied Welch's t-test and did not obtain an improvement over the difference of means. Thus, we use the difference of means distinguisher due to its simplicity.

Variance Based Distinguisher. The difference of variances distinguisher outputs the difference of variances in cycles.

$$\mathcal{D}_{vars} = \arg\max_{\hat{k}}(\text{var } \tau_{H_1} - \text{var } \tau_{H_0})$$

Note that, as before, the variance of H_0 should be smaller than that of H_1. However, outliers can badly affect this distinguisher. In cache attacks, significant outliers that can be orders of magnitude larger than regular data are not uncommon and need to be filtered to make this distinguisher work. Since H_i is key dependent, the guessed key $\hat{k}$ that maximizes the difference is the most likely to be correct. Note that the sign carries information in all three tests. In fact, the case H_0 and its leakage f_0 correspond to fewer cache misses, hence a lower miss counter, a lower average (mean) access time, and also a lower variance. The results will show that taking the sign into account derives a much better distinguisher.

When the three distinguishers are compared, the miss counter approach has the most interesting properties: It is quite intuitive, as cache misses and hits are what we are looking for. Furthermore, the method is only marginally affected by outliers. The main disadvantage of this method is the requirement of a threshold, which is processor-dependent and requires some minimal profiling. The other two methods are more affected by outliers.

All three distinguishers can easily be converted to a correlation method. Indeed, by correlating the right term (e.g. $\bar{\tau}_{H_0}$) to 0 for H_0 (a guaranteed cache hit with low reload time) and 1 for H_1 (a possible cache miss with higher reload time), the most likely key $\hat{k}$ features the highest correlation.

3.3 Attack Scenarios

Next, we describe the principles of our new *Flush+Reload* attack as well as the original and the improved versions of the attack in [15]. We will refer to the attack in [15] as the Fully Synchronous Attack (FSA) and the improved version of it with the additional AES detection step as the Semi-Synchronous Attack (SSA). Finally, the attack scenario where the attacker requires no synchronization with the server will be referred as the Asynchronous Attack (ASA). In the following, we explain and compare these attacks in detail, listing challenges and advantages of each version.

FSA. This is the original attack used in [15] where the attacker first flushes the T tables, then sends a plaintext to the encryption server to trigger an AES encryption. The server receives the plaintext from the attacker, and sends the ciphertext back. Upon receipt of the ciphertext, the attacker reloads the monitored T table blocks to learn which entries were accessed by the encryption.

SSA. In this version of the attack, we improved over the FSA by detecting the AES encryption using the *Flush+Reload* method but there is still a need for trigger event by the adversary. The advantage of this attack is the usage of an AES encryption detector that detects whether the victim is performing an AES encryption. Once the AES encryption function call is detected, the attacker flushes the monitored T table blocks **during** the AES execution in between AES rounds. Flushing in between rounds reduces the number $l - 1$ of unrelated accesses to the T table accesses, hence increasing the number of memory accesses for case H_1. In addition, we know that the detection algorithm takes half of the timing of an AES encryption. Therefore, at least half of the rounds of the AES encryption is eliminated by this detection mechanism. This results in a more biased distribution f_1, i.e. a stronger leakage. Consequently, the attack succeeds with fewer encryptions.

ASA. In the ASA, we improve over the previous two attacks by not requiring any trigger event by the adversary. Instead, plaintexts are generated by the server in regular intervals of 5M cycles. The adversary uses an AES detector to detect the AES function call and perform the *Flush+Reload* attack on the fly. In addition the network is monitored to recover transmitted ciphertexts. Unlike the previous attacks, this attack is a true ciphertext-only attack.

Note that the ASA presents a more realistic attack scenario than those presented in [12,15]. In [12] Gullasch et al. described a *Flush+Reload* attack on AES implementation of the OpenSSL library where they overload the CPU and suspend the AES encryption by controlling CFS. In [15] authors require synchronization with the server through the plaintext generation. In contrast to these previous attacks, our attack differs in the following ways:

- Our attack flushes the T tables **during** the AES encryption rather than before;
- CFS exploitation or any other type of CPU overloading is not necessary;
- Synchronization through the plaintext is no longer required, but the AES encryption call is detected instead;
- Improved side-channel data analysis/key recovery methods recover the key with fewer encryptions.

4   Experiment Setup

For the experiments, we have used the following two setups;

- **Native Execution:** In this setup, the AES encryption process and the attacker run on a native Ubuntu 12.04 LTS version with no virtualization. In this setting, we have used a two core Intel i5-2430M CPU clocked at 2.4 GHz. The purpose of this scenario is to run the attack in an environment with minimal noise and to achieve comparability to former non cross-VM cache attacks.
- **Cross-VM Execution:** In this setup, two up-to-date Ubuntu VMs, VM1 and VM2 are launched and managed by VMware ESXI 5.5 baremetal hypervisor. The attacks are then performed across hypervisor isolation boundaries.

The first VM is used as the target that does the AES encryption while the second VM acts as the attacker and executes the *Flush+Reload* attack, trying to recover the secret key. The experiments in this setting were performed on an Intel Xeon E5-2670 v2 CPU. This setup reflects a realistic attack scenario by using a modern CPU commonly used in commercial cloud systems [1,2]. In this setup, data access from the cache takes 30 cycles and the memory takes 233 cycles on average. Also in the same specification, single AES encryption without and with pre-flushed T-tables requires 257 and 659 cycles, respectively. As the Fig. 1 shows, the timing separation between the CPU cache and the main memory is clear with very few outliers. We further observe in Fig. 1 that the AES execution time changes greatly depending on whether or not the T-tables used for the encryption are loaded in the cache.

Note that all the timing measurements in the experiments are gathered using the Read Time Stamp Counter and Processor ID (RDTSCP) instruction. The usage of the RDTSCP instruction is allowed in VMware user mode, but not in KVM. Moreover, this instruction is not emulated by the VMM but executed directly, unlike other serializing instructions like CPUID used in [15]. Also, the flushing operation is performed using the Cache Line Flush (CLFLUSH) instruction.

In all experiments, one target process executes AES encryption while the attacker process tries to recover the secret key by monitoring the T-tables with the *Flush+Reload* technique. In order to clearly show the the attack success under different assumptions, we have used two distinct attack environments.

5   Results

We performed the experiments for all three attack scenarios, i.e. FSA, SSA and ASA in both native and virtualized environments. Furthermore, we analyze the timing behavior to show the improvement on the success rate by using the three different distinguishers mentioned in Sect. 3.2: the miss counter distinguisher, the difference of means distinguisher and the difference of variances distinguisher.

At first we present and compare the scores of the key guesses using the three different distinguishers in native execution in Fig. 3. The difference of means and variances distinguishers suffer more from noise due to heavy outliers stemming from different microarchitectural sources of noise. However the experiments shown in Fig. 3 were taken cutting off outliers with an outlier threshold value of 5 times the memory access time. It can be seen that for 10,000 encryptions the three distinguishers clearly maximize the score for the correct key, i.e. 180 in this case.

Then, the results of the three different attack scenarios is presented in Table 1 by comparing the ratio between cache accesses and memory accesses for cases H_0 and H_1. The precise distribution for the SSA scenario was given in Fig. 2. Recall that without noise, the ratio should be 100 %/0 % for H_0 vs. 92 %/8 % for H_1 for the FSA scenario and even more biased for the SSA scenario. The

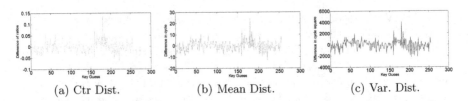

(a) Ctr Dist.   (b) Mean Dist.   (c) Var. Dist.

Fig. 3. Comparison of the scores of key guesses in the natively executed FSA scenario for three different distinguishers based on the miss counter (a), difference of means (b) and difference of variances (c), applied to 10000 traces. The correct key is 180 and clearly distinguishable in all three cases.

probability distribution shows that for H_0 approximately 95 % of the reload values are coming from L3 cache while only the 5 % come from the main memory. In H_1 however, the reload values coming from L3 cache are down to 88 %, while the values coming from the main memory increase to 12 %. Also, it can be seen from the Table 1 that there is a significant improvement in the distinguishability for SSA scenario due to flushing during AES encryption. Flushing during the encryption translates into lower noise in the T table measured access times and an improved success rate. However, the increased number of detected memory accesses for SSA is likely caused by flushes occurring *after* AES encryption has terminated. Thus, although the more realistic ASA scenario decreases the success rate, in comparison to the SSA scenario due to the difficulty of the AES detection. Hence, SSA is the most efficient way to decrease the noise and have a good resolution to find the correct key with a small number of encryptions.

Table 1. Distribution of cache accesses vs. memory accesses for the two hypotheses over the three attack scenarios. SSA provides the best distinguishability.

Attack scenarios	H_0		H_1	
	Cache	Memory	Cache	Memory
Ideal case	100 %	0 %	92 %	8 %
FSA	99 %	1 %	97 %	3 %
SSA	95 %	5 %	88 %	12 %
ASA	97 %	3 %	96 %	4 %

Finally the number of traces needed for the recovery of the key are presented in Figs. 4 and 5. As for the attack scenario success rates, our experiments in the native execution setting show that the SSA yields higher success rate than the FSA and the ASA which require 3,000, 25,000 and 30,000 encryptions, respectively. Also, the variance distinguisher works better in native setting than the other two distinguishers. For other attack scenarios e.g. the ASA, the mean distinguisher works the best, see Fig. 5(b). Note that, since ASA is the most realistic

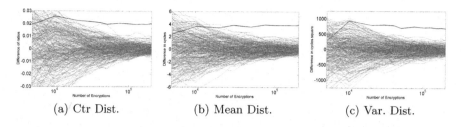

(a) Ctr Dist.   (b) Mean Dist.   (c) Var. Dist.

Fig. 4. Comparison of results in native execution for FSA scenario for different distinguishers based on the miss counter (a), difference of means (b) and difference of variances (c).

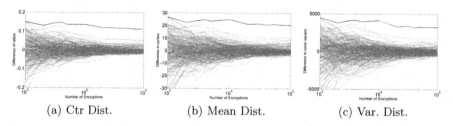

(a) Ctr Dist.   (b) Mean Dist.   (c) Var. Dist.

Fig. 5. Comparison of results in native execution for the SSA for different distinguishers based on the miss counter (a), difference of means (b) and difference of variances (c).

scenario, it requires more encryption samples than the other two, most notably compared to the SSA where only 3,000 encryption samples are needed.

5.1   Cross-VM Execution Results

In the cross-VM setting, the FSA scenario requires 30,000 encryptions to recover the full key using the miss counter hypothesis as seen in Fig. 6(a). In the same setting, 50,000 encryptions are needed when the difference of means distinguisher is used as in Fig. 6(d). As for the SSA, only 10,000 encryptions are enough to recover the full key using the mean distinguisher in Fig. 6(b). If the miss counter distinguisher is used instead of the mean distinguisher, 40,000 encryptions are needed as seen in Fig. 6(e).

For the ASA scenario, 30,000 encryptions are enough to recover the full key using miss-counter and mean distinguishers as seen in Fig. 6(c) and (f). Also, when we compare different distinguisher methods in the cross-VM setting for different attack scenarios, we see that the difference of means distinguisher works better than the miss-counter distinguisher for the most successful attack which is the SSA. While the miss-counter distinguisher gives better results for the FSA, the two distinguishers have the same impact on the results for the ASA which is the most realistic attack scenario.

We would like to note that the difference of means and the difference of variances distinguishers work better in the SSA and ASA scenarios, whereas the miss counter yields better results for the FSA. Moreover, the main advantage of using

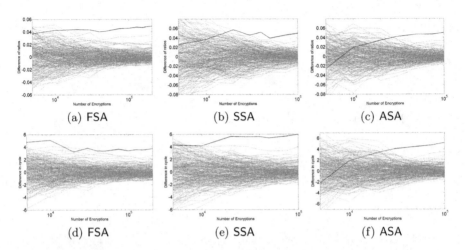

Fig. 6. Results in cross-VM execution for different attack scenarios using the miss counter distinguisher FSA (a) SSA (b) ASA (c) and the means distinguisher FSA (d) SSA (e) and ASA (f).

the variance and mean distinguishers is that they do not need an architecture dependent threshold, whereas the miss counter approach needs the access time distribution of the cache hierarchy.

Also note that the improvement of SSA is due to flushing **during** the AES execution which yields lower noise in the reloading stage. As for the ASA, we would like to emphasize that the higher number of encryptions requirement is due to the more realistic nature of the attack setting i.e. the lack of synchronization between the server and the spy process. Finally, we would like to remark that **only 15 s** are enough to recover the whole key in SSA scenario, which to the best of our knowledge is the fastest working attack in a realistic cross-VM setting without the scheduler exploitation.

6   Conclusion

In conclusion, we would like to remark that in this work, for the first time we have accomplished a cache side-channel attack on AES by flushing in between rounds. We also used an additional AES detection stage to create an asynchronous attack setting. In addition to that, we improved upon the previous work on cross-VM AES attacks by utilizing three different distinguishers for the key recovery. Finally, our experiments show that among three attack scenarios, SSA works with the minimum number of encryptions, requiring only 3,000 in the native and 10,000 in the cross-VM setting.

Acknowledgements. This work is supported by the National Science Foundation, under grant CNS-1318919.

References

1. Amazon EC2 Instances. http://aws.amazon.com/ec2/instance-types/
2. Google Compute Engine Instance Types. https://cloud.google.com/compute/docs/machine-types
3. CFS scheduler, April 2014. https://www.kernel.org/doc/Documentation/scheduler/sched-design-CFS.txt
4. Kernel Samepage Merging, April 2014. http://kernelnewbies.org/Linux_2_6_32#head-d3f32e41df508090810388a57efce73f52660ccb/
5. Acııçmez, O., Koç, Ç.K.: Trace-driven cache attacks on AES (short paper). In: Ning, P., Qing, S., Li, N. (eds.) ICICS 2006. LNCS, vol. 4307, pp. 112–121. Springer, Heidelberg (2006)
6. Benger, N., van de Pol, J., Smart, N.P., Yarom, Y.: "Ooh Aah.. Just a Little Bit" : a small amount of side channel can go a long way. In: Batina, L., Robshaw, M. (eds.) CHES 2014. LNCS, vol. 8731, pp. 75–92. Springer, Heidelberg (2014)
7. Bernstein, D.J.: Cache-timing attacks on AES (2004). http://cr.yp.to/papers.html#cachetiming
8. Bonneau, J.: Robust final-round cache-trace attacks against AES. IACR Cryptology ePrint Archive 2006/374 (2006)
9. Cai-Sen, C., Tao, W., Xiao-Cen, C., Ping, Z.: An Improved trace driven instruction cache timing attack on RSA. Cryptology ePrint Archive, Report 2011/557 (2011). http://eprint.iacr.org/
10. Labovitz, C.: How big is amazons cloud? (2012). http://www.deepfield.com/2012/04/how-big-is-amazons-cloud/
11. Dhem, J.-F., Koeune, F., Leroux, P.-A., Mestré, P., Quisquater, J.-J., Willems, J.-L.: A practical implementation of the timing attack. In: Quisquater, J.-J., Schneier, B. (eds.) CARDIS 2000. LNCS, vol. 1820, pp. 167–182. Springer, Heidelberg (2000)
12. Gullasch, D., Bangerter, E., Krenn, S.: Cache games - bringing access-based cache attacks on AES to practice. In: IEEE Symposium on Security and Privacy, pp. 490–505 (2011)
13. Hu, W.-M.: Lattice scheduling and covert channels. In: Proceedings of the 1992 IEEE Symposium on Security and Privacy, SP 1992, Washington, DC, USA, pp. 52–61. IEEE Computer Society (1992)
14. Irazoqui, G., Inci, M.S., Eisenbarth, T., Sunar, B.: Fine grain cross-VM attacks on xen and vmware are possible! Cryptology ePrint Archive, Report 2014/248 (2014). http://eprint.iacr.org/
15. Irazoqui, G., Inci, M.S., Eisenbarth, T., Sunar, B.: Wait a minute! a fast, cross-VM attack on AES. In: Stavrou, A., Bos, H., Portokalidis, G. (eds.) RAID 2014. LNCS, vol. 8688, pp. 299–319. Springer, Heidelberg (2014)
16. Jones, M.T.: Inside the linux 2.6 completely fair scheduler, December 2009. http://www.ibm.com/developerworks/library/l-completely-fair-scheduler/l-completely-fair-scheduler-pdf.pdf
17. Jones, M.T.: Anatomy of linux kernel shared memory, April 2010. http://www.ibm.com/developerworks/linux/library/l-kernel-shared-memory/l-kernel-shared-memory-pdf.pdf/
18. Kelsey, J., Schneier, B., Wagner, D., Hall, C.: Side channel cryptanalysis of product ciphers. J. Comput. Secur. **8**(2–3), 141–158 (2000)
19. Kocher, P.C.: Timing attacks on implementations of Diffie-Hellman, RSA, DSS, and other systems. In: Koblitz, N. (ed.) CRYPTO 1996. LNCS, vol. 1109, pp. 104–113. Springer, Heidelberg (1996)

20. Kocher, P.C., Jaffe, J., Jun, B.: Differential power analysis. In: Wiener, M. (ed.) CRYPTO 1999. LNCS, vol. 1666, pp. 388–397. Springer, Heidelberg (1999)
21. Mangard, S., Oswald, E., Popp, T.: Power Analysis Attacks: Revealing The Secrets of Smart Cards, vol. 31. Springer, Heidelberg (2008)
22. Osvik, D.A., Shamir, A., Tromer, E.: Cache attacks and countermeasures: the case of AES. In: Pointcheval, D. (ed.) CT-RSA 2006. LNCS, vol. 3860, pp. 1–20. Springer, Heidelberg (2006)
23. Page, D.: Theoretical use of cache memory as a cryptanalytic side-channel (2002)
24. Ristenpart, T., Tromer, E., Shacham, H., Savage, S.: Hey, you, get off of my cloud: exploring information leakage in third-party compute clouds. In: Proceedings of the 16th ACM Conference on Computer and Communications Security, CCS 2009, New York, NY, USA, pp. 199–212. ACM (2009)
25. Suzaki, K., Iijima, K., Yagi, T., Artho, C.: Memory deduplication as a threat to the guest OS. In: Proceedings of the Fourth European Workshop on System Security, p. 1. ACM (2011)
26. Tsunoo, Y., Saito, T., Suzaki, T., Shigeri, M., Miyauchi, H.: Cryptanalysis of DES implemented on computers with cache. In: Walter, C.D., Koç, Ç.K., Paar, C. (eds.) CHES 2003. LNCS, vol. 2779, pp. 62–76. Springer, Heidelberg (2003)
27. VMWare. Understanding Memory Resource Management in VMware vSphere 5.0. http://www.vmware.com/files/pdf/mem_mgmt_perf_vsphere5.pdf
28. Waldspurger, C.A.: Memory resource management in VMware ESX server. ACM SIGOPS Operating Syst. Rev. 36(SI), 181–194 (2002)
29. Yarom, Y., Falkner, K.: FLUSH+RELOAD: a high resolution, low noise, L3 cache side-channel attack. In: 23rd USENIX Security Symposium (USENIX Security 14), San Diego, CA, pp. 719–732. USENIX Association, August 2014
30. Zhang, Y., Juels, A., Oprea, A., Reiter, M.K.: HomeAlone: co-residency detection in the cloud via side-channel analysis. In: Proceedings of the 2011 IEEE Symposium on Security and Privacy, SP 2011, Washington, DC, USA, pp. 313–328. IEEE Computer Society (2011)
31. Zhang, Y., Juels, A., Reiter, M.K., Ristenpart, T.: Cross-VM side channels and their use to extract private keys. In: Proceedings of the 2012 ACM Conference on Computer and Communications Security, CCS 2012, New York, NY, USA, pp. 305–316. ACM (2012)

Faster Software for Fast Endomorphisms

Billy Bob Brumley[(✉)]

Department of Pervasive Computing,
Tampere University of Technology, Tampere, Finland
billy.brumley@tut.fi

Abstract. GLV curves (Gallant et al.) have performance advantages over standard elliptic curves, using half the number of point doublings for scalar multiplication. Despite their introduction in 2001, implementations of the GLV method have yet to permeate widespread software libraries. Furthermore, side-channel vulnerabilities, specifically cache-timing attacks, remain unpatched in the OpenSSL code base since the first attack in 2009 (Brumley and Hakala) even still after the most recent attack in 2014 (Benger et al.). This work reports on the integration of the GLV method in OpenSSL for curves from 160 to 256 bits, as well as deploying and evaluating two side-channel defenses. Performance gains are up to 51 %, and with these improvements GLV curves are now the fastest elliptic curves in OpenSSL for these bit sizes.

Keywords: Elliptic curve cryptography · GLV curves · Side-channel analysis · Timing attacks · Cache-timing attacks · OpenSSL

1 Introduction

With respect to performance, the most critical operation in an elliptic curve cryptography (ECC) implementation is scalar multiplication. The most common methods for scalar multiplication are analogous to modular exponentiation, but with signed digit sets since inverting elliptic curve points is simple. However, the number of elliptic curve point doublings in all of these methods is essentially the same, and bounded by the bit length of the scalar.

In 2001, Gallant et al. show how to halve the number of point doublings through clever choice of elliptic curve parameters [9]. These so-called "GLV curves" can exploit a fast curve endomorphism that can be computed on-the-fly and splits a single scalar multiplication into a 2-dimension multi-scalar multiplication with both arguments of roughly half the bit length.

While the theoretical gains from the GLV method are well understood, unfortunately it has yet to find its way into elliptic curve software libraries. For example, OpenSSL supports a number of GLV curves as named curves with bit sizes ranging from 160 to 256 bits but treats them as generic elliptic curves – that is, the software does not exploit the fast endomorphism for efficient scalar multiplication.

© Springer International Publishing Switzerland 2015
S. Mangard and A.Y. Poschmann (Eds.): COSADE 2015, LNCS 9064, pp. 127–140, 2015.
DOI: 10.1007/978-3-319-21476-4_9

Adding to the list of deficiencies, the elliptic curve portion of OpenSSL has known side-channel vulnerabilities. In 2009, Brumley and Hakala present the first public cache-timing attack against ECC in OpenSSL – using the vulnerability to recover a 160-bit ECC private key [5]. In 2014, Benger et al. build on that work and recover a 256-bit ECC private key with less queries [2]. The vulnerabilities remain unpatched to this day.

Motivated by these deficiencies, this work reports results of integrating the GLV method into the OpenSSL code base (Sect. 3). The performance numbers in Sect. 4 show up to a 51 % improvement. Furthermore, results of integrating two side-channel defenses show that up to 33 % improvement can be retained in tandem with the GLV method. Lastly, this work also evaluates said side-channel defenses to assess their effectiveness against data and instruction cache-timing attacks.

2 Background

This section contains background on GLV curves, OpenSSL's implementation of ECC and supported standard curves, and side-channel attacks on said implementation.

2.1 GLV Curves

The speed at which an ECC software library performs scalar multiplication is an extremely important metric. Most of the methods are some variant of a left-to-right double-and-add algorithm, perhaps with large (signed) digit sets. While the average number of point additions will vary depending on the specific method chosen, the number of point doublings is essentially the same – the bit length of the scalar.

In 2001, Gallant et al. show that clever choice of curve parameters can actually halve the number of point doublings [9]. While the authors consider a number of different curve types, their Example 4 is the most relevant to this paper [9, Sect. 2]. Let $p = 1 \mod 3$ and consider the following curve.

$$E(\mathbb{F}_p) : y^2 = x^3 + b$$

For this choice of p, there exists $\beta \in \mathbb{F}_p^*$ where $\text{ord}(\beta) = 3$. Observe $(x_1, y_1) \in E$ implies $(\beta x_1, y_1) \in E$. Denote this as a curve endomorphism $\phi : E \to E$ by $\phi : (x, y) \mapsto (\beta x, y)$. Denote $n = \#E$. Then $\phi(P) = \lambda P$ for some $\lambda \in \mathbb{F}_n^*$ where $\lambda^2 + \lambda + 1 = 0 \mod n$.

While this is all very rigorous, it is not obvious how it is at all useful for scalar multiplication. The trick is to write $k = k_1 + k_2 \lambda \mod n$ for some $k_1, k_2 \approx \sqrt{n}$. Then $kP = k_1 P + k_2 \lambda P = k_1 P + k_2 \phi(P)$ and when applying ϕ on-the-fly and with k_1, k_2 half the bit length of k, this computation takes half the doublings with a 2-dimension multi-scalar multiplication method. The authors also give an algorithm to decompose scalars accordingly [9, Sect. 4]. A number of important standards feature GLV curves, details of which will be discussed later in this paper.

The number of curves for which the GLV method applies is fairly limited. More recently, Galbraith et al. show how to apply it to larger classes of curves [8]. The key idea for these GLS curves is to work over small extension fields – otherwise the scalar multiplication methods are analogous to those used in GLV. While GLS is certainly a design trend for high-speed ECC [7], unfortunately it has not seen standards adoption yet.

2.2   ECC in OpenSSL

OpenSSL first featured support for ECC in 2005. What follows is a discussion on the range of curve support in OpenSSL, some of the internal algorithms (e.g., scalar multiplication), and the side-channel weaknesses of the library.

Standardized Curves. While the OpenSSL library has support for arbitrary elliptic curves in short Weierstrass form, the ones most commonly used are the so-called "named curves". For the purposes of this work, the most interesting standardized curves supported by OpenSSL are secp160r1, nistp192, nistp224, nistp256, secp160k1, secp192k1, secp224k1, and secp256k1. These are accordingly curves from NIST or SECG. The first four are curves over prime fields in short Weierstrass form with $A = -3$ and the latter four similar but with $A = 0$ and $p = 1 \mod 3$, i.e., GLV curves. For the NIST curves OpenSSL has dedicated code for fast modular reduction – the others it uses Montgomery reduction.

Scalar Multiplication. In OpenSSL, the low level scalar multiplication algorithm used depends on many factors. Each curve has an associated method structure, that contains function pointers to common ECC operations, one of these being a fully generalized multi-scalar multiplication:

```
int EC_POINTs_mul(const EC_GROUP *group, EC_POINT *r, const BIGNUM *n,
    size_t num, const EC_POINT *p[], const BIGNUM *m[], BN_CTX *ctx);
```

where r stores the result, n is the scalar to multiply the generator by, p is an array of num points, and m is an array of corresponding scalars to multiply said points by.

But this is just the API. The actual algorithm used varies depending on the curve by setting this function pointer when instantiating curves. For example, for all curves over binary fields the called function is an iterated implementation of Montgomery's powering ladder, in fact explicitly the algorithm given by López and Dahab [12].

For curves over prime fields and of particular interest to this work, the default method uses modified windowed Non-Adjacent Form (NAF) for scalar representation. The path through the code depends on several runtime factors. OpenSSL includes functionality for precomputing limited multiples of fixed points such as generators. The simplest case is when this precomputation is not available or not particularly helpful (e.g., no n given or num is non-zero). In this case, the algorithm is Möller's interleaved scalar multiplication [13]. The code computes NAF for each scalar, calculates small multiples for each point depending on the digit set, then proceeds MSD to LSD doubling an accumulator point at each step, then looking at each scalar's digit in that position and adding the corresponding point to the accumulator for non-zero digits. That is, with respect to the precomputation each scalar is considered independently from others and omits linear combinations of the points across scalars. To be concrete, this is the traversed code path in the following cases:

- ECDSA signature verification.
- ECDSA signature generation if no precomputation is available.
- ECDH for unknown points.
- ECDH for fixed points if no precomputation is available.

In principle, the same code applies when precomputation is available. The precomputation strategy is to reduce the number of point doublings by computed small multiples of $2^j P$ for various values of j that allow scalar NAFs to be split into smaller chunks – Möller calls this NAF splitting [14]. For example, with secp256k1 a 256-bit scalar gets split up into roughly 32 chunks with 8 digits each. Comparing the two methods, when no precomputation is available each step performs one point doubling and looks at one or two NAF digits (depending on the number of scalars involved), whereas with precomputation each step looks at a large number of digits (e.g., 32) and in total there are only a handful of point doublings (e.g., 7). To be concrete, this is the traversed code path in the following cases:

- ECDSA signature generation when precomputation is available.
- ECDH for fixed points when precomputation is available.

Implementation Attacks. OpenSSL is a popular academic target for side-channel attacks. With respect to this paper, there are two existing works that are particularly relevant.

In 2009, Brumley and Hakala show a vulnerability in OpenSSL's ECC implementation that leads to ECDSA private key recovery [5]. It is an access-driven data cache-timing attack that utilizes a local spy process that is polluting the L1 data cache in parallel (yet in a different user space) to the digital signature computation. The attack works by recovering the sequence of point doublings and additions, then filtering out very specific digital signatures for which the scalar (nonce) contains long runs of doublings, hence many consecutive zero digits in NAF representation. The authors are able to succeed in recovering an secp160r1 private key with a lattice attack after querying as few as 1 K digital signatures [5, Sect. 6]. The authors describe some countermeasures [5, Sect. 7], but there is no record of source code patches on the OpenSSL development mailing list. In fact, their "Shared Context" countermeasure was later shown to be ineffective [6].

The previous attack throws away the majority of digital signatures, side-channel trace data, and potential scalar digit information. Also the spy process targets the L1 data cache on microprocessors supporting Simultaneous Multi-Threading (SMT) through HyperThreading (HT) on Intel chips, hence does not immediately carry over to non-SMT chips. Building on this previous result, Benger et al. develop a Flush+Reload cache-timing attack in 2014 that targets the last level cache (LLC) [2]. They recover a larger private key from secp256k1 with a lattice attack after querying as few as 200 digital signatures [2, Sect. 4.1]. Furthermore, SMT or HT are not prerequisites – only a multi-core setting, hence the attack has a wider range of application. The authors describe some countermeasures [2, Sect. 5], but there is no record of source code patches on the OpenSSL development mailing list.

3   Fast and Secure Software

For elliptic curves that admit a fast endomorphism, it is clear that the GLV method provides significant performance gains. Furthermore, a number of GLV curves are already present in various standards – good examples are RFC4492[1] for TLS and the Bitcoin protocol specification[2]. Further still, widespread libraries like OpenSSL support these curves.

Yet the fact remains that the GLV method has yet to permeate these implementations. Filling the gap, This section describes integrating the GLV method in OpenSSL. At the same time, it addresses side-channel attacks by integrating two specific countermeasures in OpenSSL.

3.1   GLV in OpenSSL

OpenSSL treats GLV curves as "normal" curves and does not exploit their fast endomorphisms in any way. Having said that, there are at least two important use cases for the GLV method of scalar multiplication where OpenSSL could potentially benefit.

– For ECDH operations, splitting a single scalar multiplication into a 2-dimension multi-scalar multiplication with scalars of half the bit length (hence point doublings).
– For ECDSA signature verification, splitting a 2-dimension multi-scalar multiplication into a 4-dimension multi-scalar multiplication with scalars of half the bit length (hence point doublings).

In situations where OpenSSL has precomputation available at runtime, the GLV method is not useful because OpenSSL will already use interleaving, exploiting the endomorphism $P \mapsto 2^j P$. Indeed, the novelty of the GLV method lies in exploiting a *fast* endomorphism, i.e., one that is computationally efficient at runtime.

The OpenSSL code base has a few features that make implementing the GLV method quite modular and non-intrusive. The first is the fact that it already

---

[1] https://tools.ietf.org/html/rfc4492.
[2] https://en.bitcoin.it/wiki/Protocol_specification#Signatures.

contains a fully generalized multi-scalar multiplication algorithm. The second is the fact that the scalar multiplication method is controlled by a function pointer when instantiating the curve. With these observations, this work implements a new method in OpenSSL and assigns the function pointer for the GLV curves accordingly. Said function is essentially just a wrapper around the generalized multi-scalar multiplication algorithm – it decomposes each scalar into two scalars and applies the fast curve endomorphism to the corresponding points.

The above description is exactly how the implementation meets the two mentioned use cases. The exception is when precomputation exists – in that case, the code falls back to the default method since the GLV method is of no benefit. For example, in OpenSSL this might occur in ECDSA signature generation.

3.2   Regular Scalar Encodings

Side-channel attacks such as data and instruction cache-timing attacks can exploit implementations where the sequence of elliptic curve operations depends on the key. Ideally, as the scalar multiplication algorithm is executing it presents a consistent view through these caches that is independent of the key, i.e., the sequence of point additions and doublings is fixed regardless of the scalar. One way to do this, especially with GLV in mind, would be a multi-scalar version of Montgomery's ladder (see, e.g., Bernstein [4]) – but this would have quite a large performance penalty for OpenSSL. An ideal solution with respect to the OpenSSL code base, to retain performance and for easy integration, has the following characteristics:

– Leaves the multi-scalar multiplication algorithm largely in tact.
– Uses the same digit set as NAF.
– Does not affect the precomputation strategy.

With these goals, perhaps the most elegant solution is simply a "zero-free" scalar encoding that can serve as a drop-in replacement for OpenSSL's NAF encoding function. This work uses the "(Odd) Signed-Digit Recoding Algorithm" described by Joye and Tunstall [10, Sect. 3.2]:

> The goal is to rewrite the exponent into digits that take odd values in
> $\{-(2^w - 1), \ldots, 1, 1, \ldots, 2^w - 1\}$

and note this digit set is exactly the same as NAF but contains no zeros. The authors accomplish this intuitively by choosing signed digits such that the remaining integer to be expanded is always odd. For brevity, this work refers to this encoding as Regular NAF (RNAF). Modifying OpenSSL's multi-scalar multiplication algorithm to utilize this encoding requires only a two line change in the function, hence is minimally invasive.

3.3   Software Multiplexing

Encoding the scalar to produce a fixed sequence of point additions and doublings is enough to thwart instruction cache-timing attacks. An instruction cache-timing trace will reveal this sequence to an attacker, but it is already known a priori. On the other hand, data cache-timing attacks are still a concern. Specifically, each point addition is a table lookup where the index is a scalar digit. A data cache-timing trace can reveal these digits to an attacker.

Software multiplexing is a tool that can be leveraged to remove these traditional table lookups. In general, the approach is useful for removing any kind of conditional branch in software, including if statements and table lookups. Software multiplexing is a well-understood method to cryptographers – two examples from the literature are particularly relevant for this work.

For Curve25519 [3] finite field arithmetic, Bernstein works in an equivalence class using a representation that is not necessarily the canonical smallest non-negative residue. This allows easier modular reductions without conditional statements – better for security, better for performance to not stall the pipeline, and better for parallelization. However, at the end of scalar multiplication the resulting point must have coordinates that are the smallest non-negative residue. Bernstein does this by subtraction and building a mask from the sign, then selecting the proper value with bitwise operations.

However, table lookups slightly differ. The best example from the literature is Käsper's work for the nistp224 elliptic curve [11, Sect. 3.4]:

> We loop through the whole precomputation table in a fixed order. While the execution time is still dependent on cache behaviour, the timing variance is independent of the secret lookup index, thus leaking no valuable timing information.

That means for each table lookup, the code traverses the entire table, and the correct value extracted from the table with bitwise operations. The mask in this case gets built from the actual lookup table index. This work uses `(index^target)-1` and a signed right shift to build the mask. It is critically important to work on the data *values* and not the *pointers*.

Integrating software multiplexing into OpenSSL's generalized scalar multiplication routine is minimally intrusive. Preceding the point addition step, the select function gets called to prepare the argument for the point addition step. Then the argument for the point addition step is the output from the select function instead of the point directly from the lookup table.

4 Results

This section looks at the performance of the code after the aforementioned modifications to OpenSSL 1.0.1l. It closes with a side-channel evaluation of the code to assess the effectiveness of the countermeasures against data and instruction cache-timing attacks.

4.1 Performance

The two ECC primitives of interest here are ECDH and ECDSA. The performance numbers compare four different bit lengths ranging from 160 to 256-bit

Table 1. ECDH operations per second. The * denotes "where applicable".

Curve	Stock	GLV	GLV*+RNAF	GLV*+MUX	GLV*+RNAF+MUX
secp160r1	6824.1	—	6222.6	6171.9	6204.4 (−9.1%)
nistp192	5707.6	—	5317.4	5280.6	5198.8 (−8.9%)
nistp224	4077.2	—	3739.0	3785.5	3753.0 (−8.0%)
nistp256	3651.3	—	3296.1	3317.2	3319.5 (−9.1%)
secp160k1	6156.4	9292.0 (50.9%)	8173.8	8214.4	8175.9 (32.8%)
secp192k1	5181.2	7826.9 (51.1%)	6880.4	6864.7	6721.2 (29.7%)
secp224k1	3784.0	5527.7 (46.1%)	4955.4	5004.3	4891.6 (29.3%)
secp256k1	3265.8	4851.1 (48.5%)	4253.3	4276.7	4357.6 (33.4%)

curves, and also compares each GLV curve with a non-GLV curve for a baseline. The benchmarking environment is an Intel Core i5-4570 (Haswell-DT, 22 nm) clocked at 3.2 GHz with 16 GB memory running 64-bit Red Hat 6.6 "Santiago". The metric is operations per second, not clock cycles per operation. The reason for this is that OpenSSL's internal benchmarking does it that way, and that is what produced the performance numbers (specifically `openssl speed ecdh` and `openssl speed ecdsa`).

Key Agreement Performance. For ECDH operations, the OpenSSL `speed` utility measures the time to compute the ECDH shared secret from the private scalar and the public point. Hence it is essentially benchmarking the speed of unknown point scalar multiplication – no precomputation is available. Table 1 lists the results. The modifications to support the GLV method bring between 46 − 51% performance improvement – when comparing each GLV curve to the corresponding non-GLV curve, the former is now significantly faster in all cases. The remaining columns quantify the cost of the side-channel countermeasures, both separately and in tandem. For the non-GLV curves the cost is between 8 − 9%. One of the most interesting observations is that with the GLV and side-channel defense modifications, the GLV curves *still* outperform the stock non-GLV curves with no such defenses.

Digital Signature Performance. For ECDSA signature generation, the OpenSSL `speed` utility utilizes precomputation. Hence the code path for the GLV method will not be exercised, and the only numbers to collect are the costs of the side-channel defenses. Table 2 holds the results. In tandem, the total cost of the side-channel defenses ranges between 15 − 20%. As expected, generally each GLV curve has similar performance to the corresponding non-GLV curve.

On the other hand, for ECDSA verifications OpenSSL will not use the precomputation. Also there is no need for side-channel defenses on the verification path because all the inputs are public. Table 3 holds the results. The improvements for the GLV curves, ranging from 29 − 34%, are due to splitting the 2-dimension multi-scalar multiplication to a 4-dimension one. When comparing

Table 2. ECDSA signatures generated per second.

Curve	Stock	RNAF	MUX	RNAF+MUX
secp160r1	19256.8	16247.6	16278.3	16244.6 ($-15.6\,\%$)
nistp192	15968.0	13089.3	13099.0	13079.3 ($-18.1\,\%$)
nistp224	12954.2	10259.1	10326.3	10348.5 ($-20.1\,\%$)
nistp256	11067.5	8881.7	8896.7	8872.9 ($-19.8\,\%$)
secp160k1	18970.1	16131.7	16118.3	16121.2 ($-15.0\,\%$)
secp192k1	15629.7	12834.8	12819.6	12814.7 ($-18.0\,\%$)
secp224k1	12676.9	10441.1	10383.2	10352.5 ($-18.3\,\%$)
secp256k1	10760.9	8708.1	8687.7	8682.9 ($-19.3\,\%$)

Table 3. ECDSA signatures verified per second.

Curve	Stock	GLV
secp160r1	5526.9	—
nistp192	4623.8	—
nistp224	3353.6	—
nistp256	2912.5	—
secp160k1	5131.7	6676.5 (30.1 %)
secp192k1	4251.0	5484.0 (29.0 %)
secp224k1	3202.3	4175.2 (30.4 %)
secp256k1	2730.8	3672.9 (34.5 %)

each GLV curve to the corresponding non-GLV curve, the former is now significantly faster in all cases.

4.2 Security

The goal of this section is to provide some evidence that the side-channel defenses are effective. To this end, what follows is trace analysis for data (see [6] for the spy code) and instruction (see [1] for the spy code) cache-timing traces procured by spy processes on a microprocessor with HT. These are L1 traces for a cache with 64 sets. The spy process is executing in parallel with an OpenSSL application performing either ECDH or ECDSA signature generation for curve secp256k1. The unprotected version in the ECDH case is inclusive of the GLV method, but with no side-channel defenses. As previously discussed, GLV method does not apply to the code path for ECDSA signature generation.

ECDH Analysis. Figure 1 shows the instruction cache traces. With no defenses (top), the red annotation shows a number of point doublings while the blue annotation shows two point additions separated by a point doubling – this leak

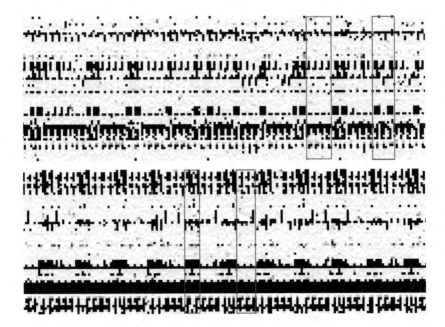

Fig. 1. ECDH through the instruction cache: with (bottom) and without (top) side-channel mitigations. y-axis: cache set index. x-axis: time. Gradient: latency from black (low) to white (high).

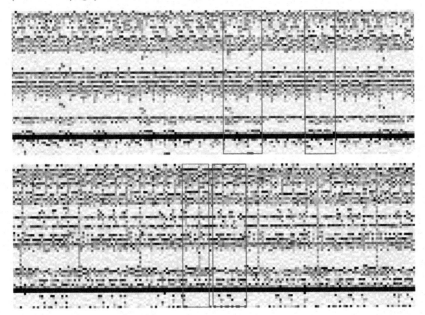

Fig. 2. ECDH through the data cache: with (bottom) and without (top) side-channel mitigations. y-axis: cache set index. x-axis: time. Gradient: latency from white (low) to black (high).

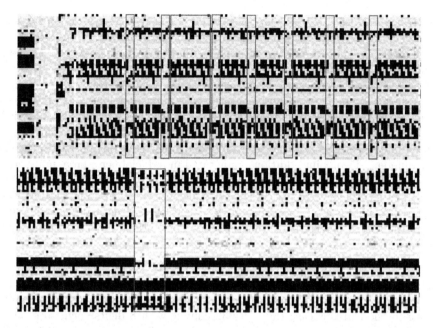

Fig. 3. ECDSA through the instruction cache: with (bottom) and without (top) side-channel mitigations. y-axis: cache set index. x-axis: time. Gradient: latency from black (low) to white (high).

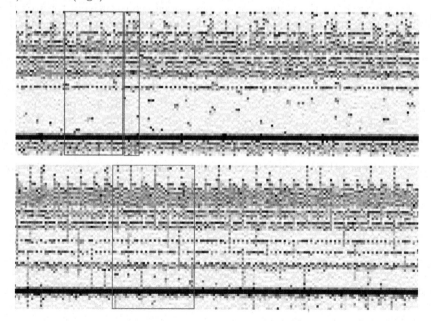

Fig. 4. ECDSA through the data cache: with (bottom) and without (top) side-channel mitigations. y-axis: cache set index. x-axis: time. Gradient: latency from white (low) to black (high).

reveals key material. With defenses (bottom), the red annotation shows two consecutive point additions while the blue annotation shows a number of point doublings. This sequence repeats throughout the trace, showing the effectiveness of the RNAF defense.

Figure 2 shows the data cache traces. With no defenses (top), the red annotation shows two point additions separated by a point doubling. The blue annotation shows two consecutive point additions. These leaks reveal key material. The trace shows the digits used in the lookup table are different because of the varying latency in many of the cache sets. With defenses (bottom), the trace is quite different. Two point additions, annotated in red, clobber a large number of cache sets. This is then followed by a number of point doublings in blue. What this suggests is the effectiveness of the MUX defense since the code traverses the entire table and has a big footprint on the cache, and furthermore the effectiveness of the RNAF defense since this sequence is essentially repeated throughout the trace.

ECDSA Analysis. Figure 3 shows the instruction cache traces. With no defenses (top), the red annotation shows all 7 point doublings – with the interleaving method in this case there are 32 chunks with 8 digits each, so exactly 7 point doublings occur. Between these, some annotated in blue, are a varying number of point additions – this leak reveals key material. With defenses (bottom), the red annotation shows four of the 7 consecutive point doublings: RNAF fixes the sequence, so the other 3 doublings appear consecutively towards the beginning of the trace (not shown). while the blue annotation shows a number of point doublings. Everything that remains is point additions. This sequence of operations reveals nothing to the attacker – the RNAF defense is working as expected.

Figure 4 shows the data cache traces. With no defenses (top), the red annotation shows one of the point doubling while the blue annotation shows a number of consecutive point additions. Observe the number of point additions between doublings varies and the latency of many of the lower cache sets reveals distinct digits used in the lookup table – this leak directly reveals key material. With defenses (bottom), the red annotation simply shows a number of point additions and highlights the fact that essentially all cache sets get clobbered as a result of the MUX defense – it is working as intended.

5 Conclusion

Using OpenSSL as a case study[3], the goal of this work is to give concrete numbers on the performance improvement realized with the GLV method, as well as to address the known side-channel vulnerabilities in OpenSSL ECC. To that end, the contributions of this work are as follows:

---

[3] http://rt.openssl.org/Ticket/Display.html?id=3667\&user=guest\&pass=guest.

- Up to 51 % performance improvement for GLV curves without side-channel defenses.
- Up to 33 % performance improvement for GLV curves with side-channel defenses.
- GLV curves *with* side-channel defenses now outperform non-GLV curves *without* side-channel defenses.
- First concrete solution (i.e., source code patch) for OpenSSL's known ECC side-channel vulnerabilities.
- Concrete evaluation of ECC software side-channel defenses, in contrast to other works that rather design for side-channel security without a platform evaluation.
- Within OpenSSL, the first application of multi-scalar multiplication for more than two scalars – better utilizing the generalized multi-scalar multiplication algorithm already present in the library.

In conclusion, this work shows that fast and secure ECC is possible for a widely-deployed software library – the concepts are not mutually exclusive.

One last subtle observation resulting from this work is that the side-channel methods to attack ECDSA depend heavily on the target application. Both published attacks [2,5] target only applications where ECDSA precomputation is *not* available – most likely not a conscience choice by the authors, but a practical difference nonetheless. This work highlights that the methods to attack applications *with* precomputation are likely very different than those without – neither of the previous attacks observe this nuance.

References

1. Acıiçmez, O., Brumley, B.B., Grabher, P.: New results on instruction cache attacks. In: Mangard, S., Standaert, F.-X. (eds.) CHES 2010. LNCS, vol. 6225, pp. 110–124. Springer, Heidelberg (2010). http://dx.doi.org/10.1007/978-3-642-15031-9_8
2. Benger, N., van de Pol, J., Smart, N.P., Yarom, Y.: "Ooh Aah.. Just a Little Bit": a small amount of side channel can go a long way. In: Batina, L., Robshaw, M. (eds.) CHES 2014. LNCS, vol. 8731, pp. 75–92. Springer, Heidelberg (2014). http://dx.doi.org/10.1007/978-3-662-44709-3_5
3. Bernstein, D.J.: Curve25519: new Diffie-Hellman speed records. In: Yung, M., Dodis, Y., Kiayias, A., Malkin, T. (eds.) PKC 2006. LNCS, vol. 3958, pp. 207–228. Springer, Heidelberg (2006). http://dx.doi.org/10.1007/11745853_14
4. Bernstein, D.J.: Differential addition chains (2006). http://cr.yp.to/ecdh/diffchain-20060219.pdf
5. Brumley, B.B., Hakala, R.M.: Cache-timing template attacks. In: Matsui, M. (ed.) ASIACRYPT 2009. LNCS, vol. 5912, pp. 667–684. Springer, Heidelberg (2009). http://dx.doi.org/10.1007/978-3-642-10366-7_39
6. Brumley, B.B., Tuveri, N.: Cache-timing attacks and shared contexts. In: Proceedings of the 2nd International Workshop on Constructive Side-Channel Analysis and Secure Design, COSADE 2011, Darmstadt, Germany, pp. 233–242, 24–25 February 2011

7. Faz-Hernández, A., Longa, P., Sánchez, A.H.: Efficient and secure algorithms for GLV-based scalar multiplication and their implementation on GLV-GLS curves (extended version). J. Cryptographic Eng. **5**(1), 31–52 (2015). http://dx.doi.org/10.1007/s13389-014-0085-7
8. Galbraith, S.D., Lin, X., Scott, M.: Endomorphisms for faster elliptic curve cryptography on a large class of curves. In: Joux, A. (ed.) EUROCRYPT 2009. LNCS, vol. 5479, pp. 518–535. Springer, Heidelberg (2009). http://dx.doi.org/10.1007/978-3-642-01001-9_30
9. Gallant, R.P., Lambert, R.J., Vanstone, S.A.: Faster point multiplication on elliptic curves with efficient endomorphisms. In: Kilian, J. (ed.) CRYPTO 2001. LNCS, vol. 2139, pp. 190–200. Springer, Heidelberg (2001). http://dx.doi.org/10.1007/3-540-44647-8_11
10. Joye, M., Tunstall, M.: Exponent recoding and regular exponentiation algorithms. In: Preneel, B. (ed.) AFRICACRYPT 2009. LNCS, vol. 5580, pp. 334–349. Springer, Heidelberg (2009). http://dx.doi.org/10.1007/978-3-642-02384-2_21
11. Käsper, E.: Fast elliptic curve cryptography in openSSL. In: Danezis, G., Dietrich, S., Sako, K. (eds.) FC 2011 Workshops. LNCS, vol. 7126, pp. 27–39. Springer, Heidelberg (2012). http://dx.doi.org/10.1007/978-3-642-29889-9_4
12. López, J., Dahab, R.: Fast multiplication on elliptic curves over $GF(2_m)$ without precomputation. In: Koç, Ç.K., Paar, C. (eds.) CHES 1999. LNCS, vol. 1717, pp. 316–327. Springer, Heidelberg (1999). http://dx.doi.org/10.1007/3-540-48059-5_27
13. Möller, B.: Algorithms for multi-exponentiation. In: Vaudenay, S., Youssef, A.M. (eds.) SAC 2001. LNCS, vol. 2259, pp. 165–180. Springer, Heidelberg (2001). http://dx.doi.org/10.1007/3-540-45537-X_13
14. Möller, B.: Improved techniques for fast exponentiation. In: Lee, P.J., Lim, C.H. (eds.) ICISC 2002. LNCS, vol. 2587, pp. 298–312. Springer, Heidelberg (2003). http://dx.doi.org/10.1007/3-540-36552-4_21

Toward Secure Implementation
of McEliece Decryption

Mariya Georgieva[1] and Frédéric de Portzamparc[1,2,3,4(✉)]

[1] Gemalto - Security Lab, Meudon, France
Mariya.Georgieva@gemalto.com
[2] INRIA, Paris-Rocquencourt Center, Paris, France
[3] Sorbonne Universités, UPMC Univ Paris 06, POLSYS, UMR 7606, LIP6,
75005 Paris, France
[4] CNRS, UMR 7606, LIP6, 75005 Paris, France
frederic.urvoy-de-portzamparc@polytechnique.org

Abstract. We analyse the security regarding timing attacks of implementations of the decryption in McEliece PKC with binary Goppa codes. First, we review and extend the existing attacks, both on the messages and on the keys. We show that, until now, no satisfactory countermeasure could erase all the timing leakages in the Extended Euclidean Algorithm (EEA) step. Then, we describe a version of the EEA never used for McEliece so far. It uses a constant number of operations for given public parameters. In particular, the operation flow does not depend on the input of the decryption, and thus closes all previous timing attacks. We end up with what should become a central tool toward a secure implementation of McEliece decryption.

1 Introduction

Context of this work. Code-based cryptography relies on the hardness of *decoding*, that is recovering $\mathbf{m}$ and $\mathbf{e}$ when given only $\mathbf{c} = \mathbf{mG} + \mathbf{e}$ and $\mathbf{G}$ (for $\mathbf{m} \in \mathbb{F}_q^k, \mathbf{G} \in \mathbb{F}_q^{k \times n}$ and $\mathbf{e} \in \mathbb{F}_q^n$). Indeed, decoding has a complexity exponential when k and the error weight grow linearly in n and no structure is known on $\mathbf{G}$ [2]. However, the error weight is critical for security for another reason: contrary to the public parameters of the code which are fixed at set by an external entity, the error may vary at each encryption, and may even be chosen by any public user (in some situations).

Therefore, a problem arises in most of the implementations of McEliece proposed (*e.g.* in [5,6,13,14]) because the operation flow of the decryption is strongly influenced by the error vector, but no information is known about the error vector when starting decryption. From an attacker's point of view, this is a favorable situation. It means that the observed or manipulated device may leak information before any detection of the attack. These security aspects were addressed by various authors, who explained that a device implementing an unprotected decryption is prone to attacks on the messages [1,12] and on the key [15,16].

© Springer International Publishing Switzerland 2015
S. Mangard and A.Y. Poschmann (Eds.): COSADE 2015, LNCS 9064, pp. 141–156, 2015.
DOI: 10.1007/978-3-319-21476-4_10

Although countermeasures were proposed against some of the leakages, the situation is still unsatisfactory, as it is noticed in the conclusion of [16]. In particular, to the best of our knowledge, no decryption algorithm requiring a number of steps independent of the error weight was described. The work of Bernstein et al. in [4] claims to achieve this goal, but some steps of the decryption (including the extended Euclidean algorithm (EEA) in the decoding) are skipped in the description, and no implementation is publicly available.

Our contributions. First, we gather the different weaknesses from [1,12,15,16]. Those attacks targeted only one of the two known methods for decoding a binary Goppa code (namely Patterson Algorithm). We evaluate how/if those threats transpose to the other decoding method (*i.e.* the alternant decoder). We detail the attacks of Strenzke and show that they can be extended to bypass the countermeasure of [15]. Our central contribution consists in describing an EEA tailored for the alternant decoder which has a flow of operations independent of the error vectors (Algorithm 8). It was inspired by a work of Berlekamp [3]. We explain step-by-step the construction of the algorithm, and provide completeness proofs (which we could not find in the literature) in the full version of this article.

2   McEliece Public-Key Encryption

We recall in Algorithm 1 the encryption and decryption in McElice PKC instantiated with a binary Goppa code, that is $q = 2$. The public key is $\mathbf{G}$ a $k \times n$ matrix over $\mathbb{F}_q$ whose rows generate a Goppa code described by the secret elements $\mathbf{x} \in \mathbb{F}_{q^m}^n$ and $g(z) \in \mathbb{F}_{q^m}[z]$ of degree t.

We detail the two possible methods for decoding a binary Goppa code. One uses the fact that Goppa codes belong to the larger class of alternant codes, so we call it the Alternant Decoder. The other one, called Patterson Algorithm, is specific to binary Goppa codes. For both, the inputs are is $\mathbf{c}$ an encoded message $\mathbf{m} \in \mathbb{F}_q^k$ with unkown error $\mathbf{e}$: $\mathbf{c} = \mathbf{mG} + \mathbf{e}$, where the Hamming weight of

---

Algorithm 1. McEliece Cryptosystem

PARAMETERS : Field size q, code length n and dimension k, parameters m, t such that $n - mt \leqslant 0$. Plaintext space: $\mathbb{F}_q^k$. Ciphertext space: $\mathbb{F}_q^n$.

KEYGEN : Pick a support $\mathbf{x} \in \mathbb{F}_{q^m}^n$, a polynomial $g \in \mathbb{F}_{q^m}[x]$ of degree t, $\mathbf{G}$ a generator matrix of $\mathscr{G}(\mathbf{x}, g)$.

PUBLIC KEY : $\mathbf{G}_{pub} = \mathbf{SGP}$, t the correction capacity of the code $\mathscr{G}(\mathbf{x}, g)$.

PRIVATE KEY : T_t a t-decoder for $\mathscr{G}(\mathbf{x}, g)$, $\mathbf{S}$ a random full rank $(n - k) \times (n - k)$ matrix , $\mathbf{P}$ a random $n \times n$ permutation matrix.

ENCRYPT :

1: Input $\mathbf{m} \in \mathbb{F}_q^k$.
2: Generate random $\mathbf{e} \in \mathbb{F}_q^n$ with $w_H(\mathbf{e}) = t$.
3: Output $\mathbf{c} = \mathbf{mG}_{pub} + \mathbf{e}$.

DECRYPT :

1: Input $\mathbf{c} \in \mathbb{F}_q^n$.
2: Compute $\tilde{\mathbf{m}} = T_t(\mathbf{cP}^{-1})$.
3: If decoding succeeds, output $\mathbf{S}^{-1}\tilde{\mathbf{m}}$, else output $\perp$.

Polynomial syndrome

$$S_{Alt,e}(z) = \sum_{\ell=0}^{2t-1} \left(\sum_{i=0}^{n-1} c_i g(x_i)^{-2} x_i^\ell \right) z^\ell.$$

Polynomials to be recovered

$$\sigma_{inv,e}(z) = \prod_{j=1}^{w} (1 - zx_{i_j}),$$

$$\omega_{inv,e}(z) = \sum_{j=1}^{w} e_{i_j} g(x_{i_j})^{-1} \prod_{\substack{s=1 \\ s \neq j}}^{w} (1 - zx_{i_s}).$$

Key equation

$(\sigma_{inv,e}, \omega_{inv,e})$ unique solution of

$$\begin{cases} \omega_{inv,e}(z) = \sigma_{inv}(z) S_{Alt,e}(z) \mod z^{2t}, \\ \deg(\sigma_{inv}) \leqslant \lfloor t/2 \rfloor, \deg(\omega_{inv}) < \lfloor t/2 \rfloor. \end{cases}$$

Resolution

$EEA(z^{2t}, S_{Alt,e}, t)$ outputs

$(\mu\sigma_{inv}, (-1)^N \mu\omega_{inv}), \mu \in \mathbb{F}_{q^m}^*, N \geqslant 0.$

Error recovery

$\sigma_e(z) = z^w \sigma_{inv}(1/z).$

Find the roots of σ_e.

Fig. 1. Alternant decoder

Polynomial syndrome

$$S_{Gop,e}(z) = \sum_{i=0}^{n-1} \frac{c_i}{z - x_i} \mod g(z).$$

Polynomials to be recovered

$$\sigma_e(z) = \prod_{j=1}^{w} (z - x_{i_j}),$$

$$\omega_e(z) = \sum_{j=1}^{w} \prod_{\substack{s=1 \\ s \neq j}}^{w} (z - x_{i_s}).$$

Key equation

(σ_1, σ_2) unique solution of

$$\begin{cases} \tau(z)\sigma_2(z) = \sigma_1(z) \mod g(z), \\ \deg(\sigma_1) \leqslant \lfloor t/2 \rfloor, \deg(\sigma_2) < \lfloor t/2 \rfloor, \\ \tau(z) = \sqrt{S_{Gop,e}(z)^{-1} + z} \mod g(z). \end{cases}$$

Resolution

1. $EEA(g(z), S_{Gop,e}(z), 0)$
   outputs $(S_{Gop,e}^{-1} \mod g)$,
2. $EEA(g(z), \tau, \lfloor t/2 \rfloor)$
   outputs (σ_1, σ_2).

Error recovery

$\sigma_e(z) = \sigma_1(z)^2 + z\sigma_2(z)^2,$

$\omega_e = \sigma_e S_e \mod g.$

Find the roots of σ_e.

Fig. 2. Patterson algorithm

$\mathbf{e}$ (denoted in the rest of this article by $w_H(\mathbf{e})$) satisfies $w_H(\mathbf{e}) \leqslant t$, and the secret $\mathbf{x}, g(z)$. The output is $\mathbf{e}$. The main steps are:

1. Compute the **polynomial syndrome** $S(z)$, a univariate polynomial deduced from $\mathbf{c}$, but depending only on $\mathbf{e}$.
2. Solve the **key equation**, which is an equation whose unkowns are univariate polynomials, using an EEA. The solutions give access to the **error locator polynomial** $\sigma_e(z)$, whose roots are related to the support elements x_{i_j} in the error positions i_j. It also the yields the **error evaluator polynomial** $\omega_e(z)$ (helpful to find the values of the errors).
3. Find the roots of $\sigma_e(z)$. Here $\mathbf{e} \in \mathbb{F}_2^n$, so $e_{i_j} \neq 0$ implies that $e_{i_j} = 1$.

The polynomial syndromes and key equations are specific to each method.

Completeness proofs are classic coding theory literature. For details, see for instance [7, Chap. 12, Sect. 9] for the Alternant Decoder and [9,18] for Patterson Algorithm.

The EEA which is used in all the available implementations (see [5,6,13,14]) consists in successive Euclidean divisions as in Algorithm 2. Its complexity is

---

Algorithm 2. Extended Euclidean Algorithm (EEA)

Input: $a(z), b(z), \deg(a) \geqslant \deg(b), d_{fin} \geqslant 0$
Output: $u(z), r(z)$ with $b(z)u(z) = r(z) \mod a(z)$ and $\deg(r) \leqslant d$

1: $r_{-1}(z) \leftarrow a(z), r_0(z) \leftarrow b(z), u_{-1}(z) \leftarrow 1, u_0(z) \leftarrow 0,$
2: $i \leftarrow 0$
3: **while** $\deg(r_i(z)) > d_{fin}$ **do**
4:     $i \leftarrow i + 1$
5:     $q_i \leftarrow r_{i-2}(z)/r_{i-1}(z)$, quotient of the Euclidean division of $r_{i-2}(z)$ by $r_{i-1}(z)$
6:     $r_i \leftarrow r_{i-2}(z) - q_i(z)r_{i-1}(z)$, rest of the Euclidean division of $r_{i-2}(z)$ by $r_{i-1}(z)$
7:     $u_i \leftarrow u_{i-2}(z) - q_i(z)u_{i-1}(z)$
8: **end while**
9: $N \leftarrow i$
10: **return** $u_N(z), r_N(z)$

$O(\deg(a)^2)$ field multiplications. Asymptotically better algorithms exist, generally referred to as Fast EEA or HGCD (for Half-GCD), with complexity $O(\deg(a) \log \deg(a))$. The reason not to use them here is that constants are hidden in the O (see for details [19]), so that for practical McEliece parameters ($t \leqslant 200$), those are not more efficient than Algorithm 2.

The EEA executions solving the key equations have complexities of $7.5 t w_H(\mathbf{e})$ (Alternant decoder) and $3.5 t w_H(\mathbf{e})$ (Patterson Algorithm), as shown in [18, Sect. 5]. The complexity of the syndrome polynomial inversion (first EEA in Patterson Algorithm) can be bounded by $2t^2$. We obtain, for a weight t error, a cost in field multiplications of $7.5t^2$ for the Alternant decoder and $5.5t^2$ for Patterson algorithm. This is why Patterson algorithm is generally preferred.

3 Decryption Oracle Attacks

3.1 Plaintext-Recovery Attacks

The attacker has a ciphertext $\mathbf{c}$ and has decryption oracle: he can request and observe the decryption of any message $\mathbf{c}' \neq \mathbf{c}$. In [1,12,15,17], the authors described attacks using the same idea (Algorithm 3).

Algorithm 3. Framework for message-recovery attacks on a decryption device.

INPUT: A valid ciphertext $\mathbf{c} = \mathbf{m}\mathbf{G}_{pub} + \mathbf{e}$, a decryption device $\mathcal{D}$.
OUTPUT: The error vector $\mathbf{e}$.

1: **for** $i = 0, \ldots, n - 1$ **do**
2:     Modify $\mathbf{c}$ into $\mathbf{c}^{\star i} = \mathbf{c} + (0, \ldots, 0, \underbrace{1}_{i-\text{th bit}}, 0, \ldots)$ and request decryption $\mathcal{D}(\mathbf{c}^{\star i})$.
3:     Deduce by timing analysis or power consumption of $\mathcal{D}$ whether $e_i = 0$ or $e_i = 1$.
4: **end for**
5: **return** Error $\mathbf{e}$.

They exploit a decryption oracle to recover the plaintext from an encrypted message $\mathbf{c}$. They focus on Patterson algorithm (Fig. 2) and propose [12, Algorithm 5, p. 171], a modified EEA which takes same execution time both on the ciphertext $\mathbf{c}$ and on the twisted $\mathbf{c}^{\star i}$.

EEA leakages in the Alternant Decoder. We adapt the framework of Algorithm 3 to the alternant decoder. The alternant decoder, as Patterson one, resorts to an EEA (Fig. 1) prone to leak information, since it execution time depends on the degree of the output, so on the error weight. Table @@reftablespsoutputspsalt gives the link of the weight of the error vector with the degree of the output of the EEA in Algorithm 1.

Table 1. Degrees of the output σ_{inv} of EEA$(z^t, S_{\mathbf{e}}(z), \lfloor t/2 \rfloor)$. ($\alpha$ denotes the position of the support such that $x_\alpha = 0$)

	$\deg(\sigma_{inv})$ if $e_\alpha = 0$	$\deg(\sigma_{inv})$ if $e_\alpha = 1$
$w_H(\mathbf{e}) = t$	t	$t-1$
$w_H(\mathbf{e}^\star) = t+1 (e_i = 0)$	t	$t-1$
$w_H(\mathbf{e}^\star) = t-1 (e_i = 1)$	$t-1$	$t-2$

After computing a polynomial $\sigma_{inv}(z)$ of degree d, if 0 belongs to the support, there are two possibilities, either the index α such that $x_\alpha = 0$ is not an error position, $\sigma_{\mathbf{e}}$ is not divisible by z, then $\deg(\sigma_{\mathbf{e}}) = \deg(\sigma_{inv})$ and $\sigma_{\mathbf{e}}(z)$ is equal to $z^{\deg(\sigma_{inv})}\sigma_{inv}(z^{-1})$, or α is an error position, and $\sigma_{\mathbf{e}}(z) = z^{\deg(\sigma_{inv})+1}\sigma_{inv}(z^{-1})$. So in this case, looking at $\deg(\sigma_{inv})$ does not distinguish manipulated ciphertexts from correct ones, and the EEA cannot be correctly protected by this method (Table 1).

Countermeasure. Building up on the countermeasure for Patterson decoding described in [12], we propose the following adaptation (Algorithm 4) to the alternant decoder. It always detects ciphertext manipulation provided that 0 is not an element of the support, and somehow restores a usual behavior of the EEA (that is, that of a valid ciphertext). The final output will not be the correct plaintext, but this is not a problem as long as the attacker cannot extract information from this result. However, we note that this protection has the same drawbacks as its Patterson equivalent: each **while** execution does not have same execution time.

Algorithm 4. Protected EEA for Alternant decoder (completes Algorithm 2)

7: $v_i \leftarrow v_{i-2}(z) - q_i(z)v_{i-1}(z)$
8: **if** $\deg(r_i) < t$ **then**
9:     Manipulate r_i so that $\deg(r_i) = \deg(r_{i-1}) - 1$ (*e.g.* $r_i \leftarrow r_i + z^{\deg(r_{i-1})-1}$).
10: **end if**

3.2   Secret Decryption Key Recovery Attacks

We address physical attacks initiated by Strenzke in [15,16] against McEliece encryption using Patterson decoding. It aims at recovering the secret key.

Generic attack scenario. The attack scenario is the following. The attacker has acces to a decryption device $\mathcal{D}$ on which he can perform physical measurements. He also knows a public encryption key, so that he can generate codewords with errors of his choice. By observing the decryption phase (more precisely, the EEA execution), Strenzke shows that one can deduce information on the support elements corresponding to the error positions. Roughly, the reason is that when a polynomial condition on those elements is satisfied, the number of iterations of the **while** loop in Algorithm 2 is reduced compared to the average number of iterations necessary to perform the EEA for error vectors of same weight. The attack consists in scanning a lot of error positions and collect sufficiently many polynomial relations so that the algebraic system obtained can be solved.

Algorithm 5 sums up the global attack framework arising from [16]. In practice, the polynomials P_w will be, for an error $\mathbf{e} = (0, \ldots, e_{i_1}, \ldots, e_{i_w}, \ldots, 0)$ with $w_H(\mathbf{e}) = w$ and $j \geqslant 0$, the evaluation of the jth elementary symmetric polynomial in w variables in $(x_{i_1}, \ldots, x_{i_w})$, that is:

$$\omega_j(\mathbf{e}) = \sum_{1 \leqslant \ell_1 < \cdots < \ell_j \leqslant w} x_{i_{\ell_1}} \ldots x_{i_{\ell_j}}.$$

Algorithm 5. Framework for key-recovery attacks on a decryption device.

INPUT: A decryption device $\mathcal{D}$, public encryption key $\mathbf{G}_{pub}$.
OUTPUT: The secret support $\mathbf{x}$.

1: **for** w well-chosen error weights **do**
2:    **for** $(i_1, \ldots, i_w)$ subset of $\{0, \ldots, n-1\}$ **do**
3:       Pick an error vector $\mathbf{e} = (0, \ldots, e_{i_1}, \ldots, e_{i_w}, \ldots, 0)$ with $w_H(\mathbf{e}) = w$.
4:       Request decryption $\mathcal{D}(\mathbf{e})$ and perform timing or power consumption analysis.

5:       **if** EEA execution faster than average (precise conditions in this Section) **then**
6:          Deduce a polynomial condition on $x_{i_1}, \ldots, x_{i_w}$ (P_w is a polynomial depending only on w):

$$P_w(x_{i_1}, \ldots, x_{i_w}) = 0 \qquad (1)$$

7:       **end if**
8:    **end for**
9: **end for**
10: Solve the non-linear system of all the collected equations (1).
11: **return** Secret support $\mathbf{x} = (x_0, \ldots, x_{n-1})$.

State-of-the-art. More precisely, Strenzke uses errors of weights $w = 1, w = 4$ and $w = 6$. For $w = 6$, errors such that Eq. (1) is satisfied are harder to find than for $w = 4$. For this reason, his strategy consists in collecting as many Eq. (1) with $w = 1$ and $w = 4$ as possible. He obtains a linear system of rank $n-m$ in the n elements of the support. Then, he selects subsets of errors of weight $w = 6$ to look for Eq. (1). These subsets are chosen so as help the polynomial system solving. According to Strenzke, for an encryption scheme with parameters $m = 10, n = 2^m, t = 40$, it takes about 15,000,000 decryption queries to collect enough equations and 28 hours to solve the algebraic system. Eventually, the full secret support $\mathbf{x}$ is recovered by the attacker, and then the Goppa polynomial is easy to find. Indeed, it is well explained in [8, p. 125] how, given the public key, it is possible to recover one from the other in polynomial time.

First Example of Leakage Exploitable by Framework 5. The first attack resorting to the method of Algorithm 5 was proposed by Strenzke in [15]. It focuses on the second EEA of Patterson Algorithm with errors of weight $w = 4$. In this case, $S_\mathbf{e}(z) = \sum_{j=1}^4 \frac{1}{z-x_{i_j}} = \frac{\omega_\mathbf{e}(z)}{\sigma_\mathbf{e}(z)}$, and

$$\omega_\mathbf{e}(z) = \underbrace{(x_{i_1} + x_{i_2} + x_{i_3} + x_{i_4})}_{\omega_1(\mathbf{e})} z^2 + \underbrace{x_{i_1} x_{i_2} x_{i_3} + x_{i_1} x_{i_2} x_{i_4} + x_{i_1} x_{i_3} x_{i_4} + x_{i_2} x_{i_3} x_{i_4}}_{\omega_3(\mathbf{e})}.$$

If $\omega_1(\mathbf{e}) = 0$, then $S_\mathbf{e}(z) = \frac{\omega_3(\mathbf{e})}{\sigma_\mathbf{e}(z)}$, and $S_\mathbf{e}^{-1} \bmod g = \omega_3(\mathbf{e})^{-1}\sigma_\mathbf{e}(z)$ therefore $\tau(z) = \sqrt{S_\mathbf{e}^{-1}(z) + z} \bmod g(z) = \sqrt{\omega_3(\mathbf{e})^{-1}\sigma_\mathbf{e}(z) + z}$ and $\tau(z)$ has degree lower than $\lfloor t/2 \rfloor$ (for $w = 4$ we have $\deg(\tau(z)) = 2$). As a consequence, the **while** test in $EEA(g(z), \tau(z), \lfloor t/2 \rfloor)$ is never fulfilled and the number of iterations N is equal to 0. When $\omega_1(\mathbf{e}) \neq 0$, $\deg(\tau(z)) > \lfloor t/2 \rfloor$ with overwhelming probability ($\tau(z)$ is a reduction modulo a polynomial of degree t), so that $N > 0$. This allows to collect many equations of the form $x_{i_1} + x_{i_2} + x_{i_3} + x_{i_4} = 0$. As Strenzke explains, the final system's rank never exceeds $n - m$. So it is not sufficient in practice to fully recover the private key. Still, he proposes a counter-measure.

Counter-measure to protect Second EEA by Strenzke. [15, Sect. 5], Strenzke proposes to detect the polynomials $\tau(z)$ leading to this leakage by checking if $\deg(\tau(z)) < \lfloor t/2 \rfloor$. This can be done just after the determination of $\tau(z)$. If so, manipulate $\tau(z)$ so that is has degree $t - 1$. This countermeasure avoids leaking information only in the second EEA, only when decoding errors of weight 4. Exploitable leakages remain, as shown in the next paragraph.

Leakage in the First EEA of Patterson Decoding. In order to complete the attack initiated in [15], Strenzke proposed in [16] to apply Algorithm 5 by focusing on time leakages in both EEA's of Patterson decoding. In [16, Corollary 1], he gives the number of iterations of the **while** loop in the first EEA. We recall it here, and complete it with the analogous result for the second EEA.

Lemma 1. *Let $\mathscr{C} = \mathscr{G}(\mathbf{x}, g(z))$ be a binary Goppa code and $S_\mathbf{e}(z)$ the polynomial syndrome associated to an error $\mathbf{e}$ with $w_H(\mathbf{e}) \leqslant \deg(g)/2 - 1$. Write $S_\mathbf{e}(z) =*

$\frac{\omega_e(z)}{\sigma_e(z)}$ mod $g(z)$. Let N_I and N_K be the number of iterations of the **while** loop respectively in $\mathrm{EEA}(g(z), S_e(z), 0)$ and $\mathrm{EEA}(g(z), \tau(z), \lfloor t/2 \rfloor)$. Then

$$N_I \leqslant \deg(\omega_e(z)) + \deg(\sigma_e(z)) \quad and \quad N_K \leqslant \deg(\omega_e(z))/2. \tag{2}$$

Proof. The result on N_I is proved in [16, Corollary 1]. Regarding N_K, observe that v_0 has degree 0 and $v_{N_K} = \sigma_2(z)$ has degree $\deg(\omega_e)/2$ (since by derivating the relation $\sigma = \sigma_1^2 + z\sigma_2^2$ we obtain $\omega_e = \sigma_2^2$). As the degrees are raised at least by one at each iteration, we obtain $N_K \leqslant \deg(\omega_e)/2$.

Errors weights $w = 4$. Pick $\mathbf{e} = (0, \ldots, e_{i_1}, \ldots, e_{i_4}, \ldots, 0)$. We have $\omega_e(z) = \omega_1(\mathbf{e})z^2 + \omega_3(\mathbf{e})$. According to Lemma 1, N_I satisfies

$$x_{i_1} + x_{i_2} + x_{i_3} + x_{i_4} \neq 0 \implies N_I = 6,$$
$$x_{i_1} + x_{i_2} + x_{i_3} + x_{i_4} = 0 \implies N_I = 4.$$

Therefore, even if the second EEA has been protected with Strenzke's counter-measure, errors of weight $w = 4$ leak the same information in the first EEA. Other equations are found by using error with weight $w = 6$.

Error weights $w = 6$. For $\mathbf{e} = (0, \ldots, e_{i_1}, \ldots, e_{i_6}, \ldots, 0)$, we have for $S_{Gop,\mathbf{e}}(z)$:

$$S_{Gop,\mathbf{e}}(z) = \frac{\omega_1(\mathbf{e})z^4 + \omega_3(\mathbf{e})z^2 + \omega_5(\mathbf{e})}{\sigma_e(z)}.$$

Strenzke's purpose is to detect for which $\mathbf{e}$ is holds that $\omega_3(\mathbf{e}) = \omega_1(\mathbf{e}) = 0$. These cases are exactly those with $S_e(z)^{-1} = \omega_5(\mathbf{e})^{-1}\sigma_e(z)$ and hence $\deg(\tau(z)) < \lfloor t/2 \rfloor$, so that $N_K = 0$ provided that Strenzke's counter-measure is not applied. This is a somehow surprising proposition, since this criterion can be rendered useless by a counter-measure already proposed by the same author.

Table 2. Overview of small- error-weight message attacks. Cases marked with a * or a $^+$ are proposed resp. in [15,16].

		$\mathrm{EEA}(g, S_e, 0)$	$\mathrm{EEA}(g, \tau, \lfloor t/2 \rfloor)$
$w_H(\mathbf{e}) = 4$	$\omega_1(\mathbf{e}) \neq 0$	$N_I \leqslant 6$	$N_K \leqslant 1$
	$\omega_1(\mathbf{e}) = 0$	$N_I \leqslant 4^*$	$N_K = 0^+$ CM $\deg(\tau) < \lfloor t/2 \rfloor^+$
$w_H(\mathbf{e}) = 6$	$\omega_1(\mathbf{e}) \neq 0, \omega_3(\mathbf{e}) \neq 0$	$N_I \leqslant 10$	$N_K \leqslant 2$
	$\omega_1(\mathbf{e}) = 0, \omega_3(\mathbf{e}) \neq 0$	$N_I \leqslant 8$	$N_K \leqslant 1$
	$\omega_1(\mathbf{e}) = 0, \omega_3(\mathbf{e}) = 0$	$N_I \leqslant 6$	$N_K = 0^*$ CM $\deg(\tau) < \lfloor t/2 \rfloor$
$w_H(\mathbf{e}) = 2w'$	$\omega_1(\mathbf{e}) \neq 0, \omega_3(\mathbf{e}) \neq 0$	$N_I \leqslant 4w' - 2$	$N_K \leqslant w' - 1$
	$\omega_1(\mathbf{e}) = 0, \omega_3(\mathbf{e}) \neq 0$	$N_I \leqslant 4w' - 4$	$N_K \leqslant w' - 2$
	$\omega_1(\mathbf{e}) = 0, \omega_3(\mathbf{e}) = 0$	$N_I \leqslant 4w' - 6$	$N_K \leqslant w' - 3$

Combination of First and Second EEA. When using error weights $w \geqslant 6$, the attacker will encounter problems due to the fact that all the values given in Table 2 (on page 7) are only bounds (except in the cases $N \leqslant 0$). Indeed, it may happen that one of the Euclidean divisions entails a degree fall greater than 1 independantly of the degree of w_e. For example, with $w = 6$, the attacker may observe $N_K = 1$ whereas $w_1(e)$ is not zero. This remark leads Strenzke to discard those cases for an attack as long as no way of distinguishing thoses cases is found. We propose such distinguisher, by using N_I to determine if $w_1(e)$ is zero, as $w_1(e) = 0$ implies $N_I \leqslant 8$. Indeed, an attacker observing the errors e with $(N_I, N_K) = (10, 1)$ can conclude that $w_1(e) \neq 0$ (Table 2). We may have $(N_I, N_K) = (8, 1)$ when $w_1(e) \neq 0$ if three cancellations occur in the 12 intermediate polynomials, which has probability $p_3 = \binom{12}{3} 2^{-3m} (1 - 2^{-m})^9 \approx 2.10^{-7}$ for $m = 10$ (we model the leading coefficients as random elements of $\mathbb{F}_{2^m}$). When sampling x error vectors, we expect to find $p_3 x$ such misleading cases. With the numbers of samples from [16, Table 2], the probability to find one is not negligible. If at least one wrong equation is deduced, the system to solve has no solution and the attack fails. We propose to avoid this problem by using errors with $w \geqslant 8$.

Error weights $w = 8$. We sampled randomly 10,000,000 errors e of weight 8 and collected the couples (N_I, N_K) in Table 3. When $w_H(e) = 8$, there are more possibilities than with $w = 6$. Samples with $(N_I \leqslant 12, N_K \leqslant 2)$ do not necessarily have $w_1(e) = 0$: this happens with probability $p_3' = \binom{17}{3} 2^{-3m} (1 - 2^{-m})^{14} \approx 6.10^{-7}$ for $m = 10$ (we found 3). In particular, the case marked with a * in Table 3 would make the attacker to think erroneously that the corresponding error vector satisfies $w_1(e) = 0$. However, the number of parasitic cancellations necessary to provide values (N_I, N_K) compatible with $(w_1(e), w_3(e)) = (0, 0)$ is 6, which happens with probability $p_3' = \binom{17}{6} 2^{-6m} (1 - 2^{-m})^{11} \approx 10^{-14}$ for $m = 10$. If $w_1(e) = 0$ but $w_3(e) \neq 0$, then a couple $(10, 1)$ is found if 3 cancellations occur. This has probability $2^{-m} p_3' \approx 6.10^{-10}$ (as w_1 takes all the values of $\mathbb{F}_{2^m}$

Table 3. Number of samples for each (N_I, N_K) for 10,000,000 error vectors with $w = 8$. Code parameters: $m = 10, n = 2^m, t = 40$. See text for explanation on *.

	No parasitic cancellation	1 parasitic cancellation	2 parasitic cancellations	3 parasitic cancellations
$w_1(e) \neq 0$ $w_3(e) \neq 0$	(14,3): 9855087	(13,3): 115439 (14,2): 18916	(12,3): 614 (13,2): 248 (14,1): 8	(12,2): 1 * (11,3): 2
$w_1(e) = 0$ $w_3(e) \neq 0$	(12,2): 9570	(11,2): 96 (12,1): 8	(10,2): 0 (11,1): 0	
$w_1(e) = 0$ $w_3(e) = 0$	(10,1): 10	(9,1): 0	(8,1): 0	

with same probability). Therefore, we are able to say without ambiguity when $(\omega_1(\mathbf{e}), \omega_3(\mathbf{e})) = (0, 0)$ on a considerable amount of samples. We deduce from our samples 10 equations $\omega_1(\mathbf{e}) = 0$ which are correct with proba. $(1 - 10^{-7})$ and 10 equations $\omega_3(\mathbf{e}) = 0$ correct with proba. $(1 - 10^{-3})$. To conclude, although our method requires more samples than the previous one (around 10^9 to collect some thousands equations with ω_1, and dozens with ω_3), we can recover information on the support even if the countermeasure $\deg(\tau) < \lfloor t/2 \rfloor$ is implemented.

Small Weight Error Messages in Alternant Decoder. We determine if an attacker can retrieve any information by applying Algorithm 5 if the Alternant decoder is implemented. First, we give in Lemma 2 the analogous of Lemma 1.

Lemma 2. *Let* $\mathbf{e}$ *be an error with* $w_H(\mathbf{e}) \leqslant t$. *Then* $S_{Alt,\mathbf{e}}(z) = \frac{\omega_{inv,\mathbf{e}}(z)}{\sigma_{inv,\mathbf{e}}(z)}$ *mod* z^{2t} *and the number of iterations* N *of the* **while** *loop of the Alternant decoder in the EEA satisfies*

$$N \leqslant N_{max} = \min(\deg(\sigma_{inv,\mathbf{e}}), \deg(S_{Alt,\mathbf{e}}) - \deg(\omega_{inv})). \tag{3}$$

Specific case of weight 1 errors. If $w = 1$, we always have $\deg(\omega_{inv}) = 0$ and $\deg(\sigma_{inv}) = 1$ except if $x_{i_1} = 0$. Indeed, in this case, the polynomial syndrome is a constant: $S_{\mathbf{e}}(z) = \frac{1}{g(0)^2}$ and the **while** loop is never executed (Table 3).

Error weights $w > 1$. We suppose that no error occurred in the zero element of the support so that $\deg(\sigma_{inv}) = w_H(\mathbf{e})$ always holds (the coefficient of z^w in σ_{inv} is $x_{i_1} \ldots x_{i_w}$). Therefore, faster decryptions indicate the cancellation of a leading coefficient in the intermediate values, but in the alternant decoder we found no way of determining which intermediate value was concerned. If by any chance a power analysis can ensure that it is the first intermediate polynomial (that is, the syndrome polynomial $S_{Alt,\mathbf{e}}(z)$) that has a degree smaller than expected, then the information recovered would be:

$$\sum_{j=1}^{w} g(x_{i_j})^{-2} \sum_{j=1}^{w} x_{i_j}^{2t-1} = 0. \tag{4}$$

We observe that the equations written thanks to this method are more complex than with Patterson algorithm, at least for two reasons. First, they are not directly polynomial, and the degrees implied are much higher. Second, as both $\mathbf{x}$ and g have to be unknown ([8, p. 125]), additive unknowns are necessary: either $t + 1$ to describe the secret polynomial's coefficients, or n if we introduce new equations $y_i = g(x_i)^{-2}$. We conclude that the alternant decoder is intrinsically more resistant to Strenzke's attacks. However, the overall security is still not clear due to the uncertainty on the countermeasure (Algorithm 4) against Algorithm 3.

4   Extended Euclidean Algorithm with Constant Flow

We expose a way of implementing the EEA algorithm unused so far for McEliece decryption. It has the very interesting property of requiring a number of operations depending only on the Goppa polynomial degree t and **not** on the weight of the error introduced in the ciphertext. Therefore, the attacks of Sects. 3.1 and 3.2 are not possible.

It is inspired by Berlekamp's work in [3] (which as followed by other works of optimization in the VLSI community, amongst many others [10,11]). We could find no reference to it in any paper related to McEliece. On the contrary, designing such an algorithm is desirable goal according to the conclusion of [16]. The reason may be that [3] has a very limited access, and we could find no completeness proofs of the algorithm proposed. Here, we transform smoothly the original EEA (Algorithm 2) into successive version gaining in regularity (Algorithms 6 and 7). We end up with Algorithm 8, which is simpler and more regular than all the previous ones. At each step, we give and prove (in the full version of this article) the form of the outputs and intermediate values. Finally, each execution of Algorithm 8 costs, in field multiplications, exactly $16t^2$ ($2t$ times a loop costing $4 \times 2t$).

In the rest of this article we will set N be the number of Euclidean divisions performed during $\mathrm{EEA}(z^{2t}, S_{Alt}(z), t)$ in Algorithm 2, $d_i = \deg(r_i(z))$, and $\delta_i = \deg(q_i(z)) = \deg(r_{i-2}) - \deg(r_{i-1})$. For any polynomial $P(z) \in \mathbb{F}_{q^m}[z]$, we denote its coefficients by P_j even for $j > \deg(P)$ (in which case $P_j = 0$), so that

$$P(z) = \sum_{j=0}^{+\infty} P_j z^j = P_{\deg(P)} z^{\deg(P)} + \cdots + P_0.$$

Regarding the δ_i's, we have:

$$\sum_{i=1}^{N} \delta_i = \deg(u_N(z)) = \deg(\omega_{inv,\mathbf{e}}) = w_H(\mathbf{e}) - 1.$$

Unrolling Euclidean Divisions. In Algorithm 6, we decompose each Euclidean division into a number of polynomial subtractions depending only on δ_i the degrees of the quotients. We explicit the intermediate values of the Euclidean division of $R_{i-2}(z)$ by $R_{i-1}(z)$, that we denote by $R_i^{(0)}(z), \ldots, R_i^{(\delta_i+1)}(z)$. To do so, we eliminate in each $R_i^{(j)}(z)$ (for $0 \leqslant j \leqslant \delta_i + 1$) the term $z^{d_{i-2}-j}$, whether the associated coefficient is zero or not. This is why we perform the Euclidean divisions in a way to avoid the divisions by a field elements (Steps 7 to 10 of Algorithm 6). Consequently, the outputs are multiple of the outputs of Algorithm 2.

Proposition 1 (Comparison of Algorithms 2 and 6). *Let $a(z)$ and $b(z)$ be two polynomials with $\deg(a(z)) \geqslant \deg(b(z))$, and d a non-negative integer. $u_i(z), v_i(z), r_i(z), q_i(z)$ are the intermediate values in Algorithm 2, and*

Algorithm 6. EEA with unrolled Euclidean Division

Input: $a(z) = z^{2t}, b(z) = S_\mathbf{e}(z)$.
Output: $U(z) = \lambda_N \sigma_\mathbf{e}(z), R(z) = \lambda_N \omega_\mathbf{e}(z)$ (for some $\lambda_N \in \mathbb{F}_{q^m}^*$).

1: $R_{-1}(z) \leftarrow a(z), R_0(z) \leftarrow b(z), U_{-1}(z) \leftarrow 1, U_0(z) \leftarrow 0, i \leftarrow 0$.
2: **while** $\deg(R_i(z)) > t$ **do**
3:      $i \leftarrow i + 1$
4:      $R_{i-2}^{(0)}(z) \leftarrow R_{i-2}(z), U_{i-2}^{(0)}(z) \leftarrow U_{i-2}(z)$
5:      $\Delta_i \leftarrow \deg(R_{i-2}) - \deg(R_{i-1})$
6:      $\beta_i \leftarrow \mathrm{LC}(R_{i-1}(z))$.
7:      **for** $j = 0, \ldots, \Delta_i$ **do**
8:          $\alpha_{i,j} \leftarrow R_{i,d_{i-2}-j}^{(j)}$,
9:          $R_{i-2}^{(j+1)}(z) \leftarrow \beta_i R_{i-2}^{(j)}(z) - \alpha_{i,j} z^{\Delta_i - j} R_{i-1}(z)$
10:         $U_{i-2}^{(j+1)}(z) \leftarrow \beta_i U_{i-2}^{(j)}(z) - \alpha_{i,j} z^{\Delta_i - j} U_{i-1}(z)$
11:     **end for**
12:     $R_i(z) \leftarrow R_{i-2}^{(\Delta_i+1)}(z), U_i(z) \leftarrow U_{i-2}^{(\Delta_i+1)}(z)$
13: **end while**
14: $N \leftarrow i$.
15: **return** $U_N(z), R_N(z)$

$U_i(z), V_i(z), R_i(z)$ *are the intermediate values in Algorithm 6. It holds that, for all* $i = -1, \ldots, N$, *there exists* $\lambda_i \in \mathbb{F}_{q^m}^*$ *such that:*

$$R_i(z) = \lambda_i r_i(z),$$
$$U_i(z) = \lambda_i u_i(z).$$

Hence, $\Delta_i = \deg(R_{i-2}) - \deg(R_{i-1}) = \deg(r_{i-2}) - \deg(r_{i-1}) = \delta_i$ *for all* i.

There are two problems with Algorithm 6. The first one is that the inner **for** loop (Steps 7 to 11) has a variable length, and contains a multiplication $z^{\delta_i-(j-1)} R_i(z)$ which depends on the iteration, which will produce a recognizable pattern. The second problem is that the **while** loop leads to a number of operations depending on the input. Algorithm 7 is a first step towards the resolution of the second problem. It is not realistic (it requires to know the δ_i's), but it eases the proofs of completeness of Algorithm 8, which solves both issues.

Regular Polynomial Shift Pattern. In Algorithm 7, we perform the Euclidean division in such a way that we only multiply the operand by z at each **for** iteration. This can be done by splitting in two phases each Euclidean divisions. The first phase (Steps 4 to 7) "re-aligns" the operands $\tilde{R}_{i-2}$ and $\tilde{R}_{i-1}$ so that they both have same degree $d = \deg(R_{-1}(z))(= 2t)$. Doing so, the second phase (Steps 8 to 12) compute the polynomial subtractions (corresponding to Steps 9-10 of Algorithm 6) and perform a shift "re-aligning" the operands. A consequence is that the polynomials $\tilde{R}_i(z)$ are of the form $z^{k_i} R_i(z)$ and the degrees d_i are lost.

Algorithm 7. Toy EEA with regular shift pattern

Input: $a(z) = z^{2t}, b(z) = S_{\mathbf{e}}(z), d = 2t$
Output: $\tilde{U}_N(z) = z^{d-d_N-1+1}\sigma_{\mathbf{e}}(z), \tilde{R}_N(z) = z^{d-d_N-1+1}\omega_{\mathbf{e}}(z).$

1: $\tilde{R}_{-1}(z) \leftarrow a(z), \tilde{R}_0(z) \leftarrow zb(z), \tilde{U}_{-1}(z) \leftarrow 1, \tilde{U}_0(z) \leftarrow 0.$
2: **for** $i = 1, \ldots, N$ **do**
3:     $\tilde{R}_{i-2}^{(0)}(z) \leftarrow \tilde{R}_{i-2}(z), \tilde{U}_{i-2}^{(0)}(z) \leftarrow \tilde{U}_{i-2}(z)$
4:     **for** $j = 1, \ldots, \Delta_i - 1$ **do** $\left.\vphantom{\begin{array}{c} \\ \\ \end{array}}\right\} L_1$
5:        $\tilde{R}_{i-1}(z) \leftarrow z\tilde{R}_{i-1}(z)$
6:        $\tilde{U}_{i-1}(z) \leftarrow z\tilde{U}_{i-1}(z)$
7:     **end for**
8:     **for** $j = 0, \ldots, \Delta_i$ **do** $\left.\vphantom{\begin{array}{c} \\ \\ \\ \\ \\ \end{array}}\right\}$
9:        $\tilde{\alpha}_{i,j} \leftarrow \tilde{R}_{i,d}^{(j)}, \tilde{\beta}_i \leftarrow \tilde{R}_{i-1,d}.$
10:       $\tilde{R}_{i-2}^{(j+1)}(z) \leftarrow z\left(\tilde{\beta}_i\tilde{R}_{i-2}^{(j)}(z) - \tilde{\alpha}_{i,j}\tilde{R}_{i-1}(z)\right)$ $\left.\vphantom{\begin{array}{c} \\ \\ \\ \end{array}}\right\} L_2$
11:       $\tilde{U}_{i-2}^{(j+1)}(z) \leftarrow z\left(\tilde{\beta}_i\tilde{U}_{i-2}^{(j)}(z) - \tilde{\alpha}_{i,j}\tilde{U}_{i-1}(z)\right)$
12:    **end for**
13:    $\tilde{R}_i(z) \leftarrow \tilde{R}_{i-2}^{(\Delta_i+1)}(z), \tilde{U}_i(z) \leftarrow \tilde{U}_{i-2}^{(\Delta_i+1)}(z)$
14: **end for**
15: **return** $\tilde{U}_N(z), \tilde{R}_N(z)$

Proposition 2 (Comparison of Algorithms 6 and 7). *For each* $i = 1, \ldots, N$, *after Step 13 of Algorithm 7, it holds that*

$$(\tilde{R}_{i-1}(z), \tilde{R}_i(z)) = (z^{d-d_{i-1}}R_{i-1}(z), z^{d-d_{i-1}+1}R_i(z)),$$
$$(\tilde{U}_{i-1}(z), \tilde{U}_i(z)) = (z^{d-d_{i-1}}U_{i-1}(z), z^{d-d_{i-1}+1}U_i(z)).$$

Complete Regular Flow EEA. To design a real constant flow algorithm, we merge the loops L_1 and L_2 in a common pattern so as to be indistinguishable (Steps 5 to 7 of Algorithm 8). They differenciate by the assignements which are performed in Steps 14-15 and 18-19. To know when polynomials substractions have to be stopped, we collect in a counter δ the number of shifts necessary to re-align the operands. Finally, when the polynomials σ_{inv} and ω_{inv} have been computed, the extra executions of the main loop (Steps 4 to 22) consist in shifting the operands. therefore, the number of iterations can be safely set to the maximum value (*i.e.* $2t$ to decode the errors with $w_H(\mathbf{e}) = t$)), and the **while** loop is replaced by **for**.

Proposition 3 (Comparison of Algorithms 6 and 8). *For each* $i = 1, \ldots, N$, *after steps 21, it holds that:*

$$\hat{R}_{2(\delta_1+\cdots+\delta_i)}(z) = z^{d-d_i-1+1}R_i(z),$$
$$\hat{U}_{2(\delta_1+\cdots+\delta_i)}(z) = z^{d-d_i-1+1}U_i(z).$$

The outputs of Algorithm 8 are, for some $\mu \in \mathbb{F}_{q^m}^*$:

$$\hat{R}_d(z) = z^{d-w_H(\mathbf{e})+1}R_N(z) = \mu z^{d-w_H(\mathbf{e})+1}\omega_{inv}(z),$$
$$\hat{U}_d(z) = z^{d-w_H(\mathbf{e})+1}U_N(z) = \mu z^{d-w_H(\mathbf{e})+1}\sigma_{inv}(z).$$

Algorithm 8. EEA with regular flow

Input: $a(z) = z^{2t}, b(z) = S_{\mathbf{e}}(z), d = 2t$
Output: $\hat{U}_d(z) = \mu z^{d-w_H(\mathbf{e})+1}\sigma_{inv}(z), \hat{R}_d(z) = \mu z^{d-w_H(\mathbf{e})+1}w_{\mathbf{e}}(z)$ for some $\mu \in \mathbb{F}_{q^m}^*$.

1: $\hat{R}_{-1}(z) \leftarrow a(z), \hat{R}_0(z) \leftarrow zb(z),$
2: $\hat{U}_{-1}(z) \leftarrow 1, \hat{U}_0(z) \leftarrow 0,$
3: $\delta \leftarrow -1.$
4: **for** $j = 1, \ldots, d$ **do**
5:     $\alpha_j \leftarrow \hat{R}_{j-1,d}, \beta_j \leftarrow \hat{R}_{j-2,d}.$
6:     $temp_R(z) \leftarrow z\left(\alpha_j\hat{R}_{j-2}(z) - \beta_j\hat{R}_{j-1}(z)\right).$
7:     $temp_U(z) \leftarrow z\left(\alpha_j\hat{U}_{j-2}(z) - \beta_j\hat{U}_{j-1}(z)\right).$
8:     **if** $\alpha_j = 0$ (i.e. $\deg(\hat{R}_{j-1}) < \deg(\hat{R}_{j-2})$) **then**
9:         $\delta \leftarrow \delta + 1.$
10:    **else**
11:        $\delta \leftarrow \delta - 1.$
12:    **end if** $\Big\} L$
13:    **if** $\delta < 0$ **then**
14:        $(\hat{R}_j(z), \hat{R}_{j-1}(z)) \leftarrow (\hat{R}_{j-1}(z), temp_R)$
15:        $(\hat{U}_j(z), \hat{U}_{j-1}(z)) \leftarrow (\hat{U}_{j-1}(z), temp_U)$
16:        $\delta \leftarrow 0.$
17:    **else**
18:        $(\hat{R}_j(z), \hat{R}_{j-1}(z)) \leftarrow (temp_R, \hat{R}_{j-2}(z))$
19:        $(\hat{U}_j(z), \hat{U}_{j-1}(z)) \leftarrow (temp_U, \hat{U}_{j-2}(z))$
20:        $\delta \leftarrow \delta.$
21:    **end if**
22: **end for**
23: **return** $\hat{U}_d(z), \hat{R}_d(z)$

Therefore, provided 0 is not an element of $\mathbf{x}$, $\hat{U}_d(z)$ allows to recover the error positions without ambiguity. Transposing this result to Patterson decoding requires to adapt both EEA's. The adaptation of the second one is straightforward. For the first one (syndrome inversion), a problem arises: the analogous of Proposition 3 would yield $\hat{U}_{N_I}(z) = \mu z^{k_i}(S_{Gop,\mathbf{e}}^{-1} \bmod g)$ for some $k_i > 0$, and we found no way of determining when z is a factor of $S_{Gop,\mathbf{e}}^{-1} \bmod g$. However, we can protect the second EEA to avoid the attack of 3.2.

5   Conclusion

We proposed an algorithm determining the error-locator polynomial costing always $16t^2$ field multiplications on any input. It contains a test depending on secret data, followed by two balanced branches. The indistinguishability of those branches by an attacker is crucial for the security of the decryption, and depends on the architecture of the implementation.

Acknowledgements. We are thankful to J.-C. Faugère, A. Gouget, L. Perret and anonymous reviewers for their comments in the preparation of this paper.

References

1. Avanzi, R., Hoerder, S., Page, D., Tunstall, M.: Side-channel attacks on the McEliece and Niederreiter public-key cryptosystems. J. Cryptographic Eng. **1**(4), 271–281 (2011)
2. Berlekamp, E., McEliece, R., van Tilborg, H.: On the inherent intractability of certain coding problems. IEEE Trans. Inf. Theory **24**(3), 384–386 (1978)
3. Berlekamp, E., Seroussi, G., Po, T.: A hypersystolic reed–solomon decoder. In: Bhargava, V.K., Wicker, S.B., IEEE Communications Society, IEEE Information Theory Society (eds.) Reed-Solomon Codes and Their Applications, pp. 205–241. IEEE Press, Piscataway (1994)
4. Bernstein, D.J., Chou, T., Schwabe, P.: McBits: fast constant-time code-based cryptography. In: Bertoni, G., Coron, J.-S. (eds.) CHES 2013. LNCS, vol. 8086, pp. 250–272. Springer, Heidelberg (2013)
5. Biswas, B.: Aspects de mise en oeuvre de la cryptographie basée sur les codes. Theses, Ecole Polytechnique X, October 2010
6. Heyse, S.: Implementation of McEliece based on quasi-dyadic goppa codes for embedded devices. In: Yang, B.-Y. (ed.) PQCrypto 2011. LNCS, vol. 7071, pp. 143–162. Springer, Heidelberg (2011)
7. MacWilliams, F.J., Sloane, N.J.A.: The Theory of Error-Correcting Codes, 5th edn. North-Holland, Amsterdam (1986)
8. Overbeck, R., Sendrier, N.: Code-based cryptography. In: Bernstein, D., Buchmann, J., Dahmen, E. (eds.) Post-Quantum Cryptography, pp. 95–145. Springer, Heidelberg (2009)
9. Patterson, N.: The algebraic decoding of Goppa codes. IEEE Trans. Inf. Theory **21**(2), 203–207 (1975)
10. Sarwate, D., Shanbhag, N.: High-speed architectures for Reed-Solomon decoders. IEEE Trans. Very Large Scale Integr. (VLSI) Syst. **9**(5), 641–655 (2001)
11. Sarwate, D., Yan, Z.: Modified Euclidean algorithms for decoding Reed-Solomon codes. In: IEEE International Symposium on Information Theory, ISIT 2009, pp. 1398–1402, June 2009
12. Shoufan, A., Strenzke, F., Molter, H.G., Stöttinger, M.: A timing attack against patterson algorithm in the McEliece PKC. In: Lee, D., Hong, S. (eds.) ICISC 2009. LNCS, vol. 5984, pp. 161–175. Springer, Heidelberg (2010)
13. Shoufan, A., Wink, T., Molter, H., Huss, S., Kohnert, E.: A Novel cryptoprocessor architecture for the McEliece public-key cryptosystem. IEEE Trans. Comput. **59**(11), 1533–1546 (2010)
14. Strenzke, F.: A smart card implementation of the McEliece PKC. In: Samarati, P., Tunstall, M., Posegga, J., Markantonakis, K., Sauveron, D. (eds.) WISTP 2010. LNCS, vol. 6033, pp. 47–59. Springer, Heidelberg (2010)
15. Strenzke, F.: A timing attack against the secret permutation in the McEliece PKC. In: Sendrier, N. (ed.) PQCrypto 2010. LNCS, vol. 6061, pp. 95–107. Springer, Heidelberg (2010)
16. Strenzke, F.: Timing attacks against the syndrome inversion in Code-Based cryptosystems. In: Gaborit, P. (ed.) PQCrypto 2013. LNCS, vol. 7932, pp. 217–230. Springer, Heidelberg (2013)

17. Strenzke, F., Tews, E., Molter, H.G., Overbeck, R., Shoufan, A.: Side channels in the McEliece PKC. In: Buchmann, J., Ding, J. (eds.) PQCrypto 2008. LNCS, vol. 5299, pp. 216–229. Springer, Heidelberg (2008)
18. Sugiyama, Y., Kasahara, M., Hirasawa, S., Namekawa, T.: A method for solving key equation for decoding goppa codes. Inf. Control **27**(1), 87–99 (1975)
19. Thull, K., Yap, C.: A Unified Approach to HGCD Algorithms for polynomials and integers (1990)

Fault Attacks

Fault Injection with a New Flavor: Memetic Algorithms Make a Difference

Stjepan Picek[1], Lejla Batina[2], Pieter Buzing[3], and Domagoj Jakobovic[1]($\boxtimes$)

[1] Faculty of Electrical Engineering and Computing, University of Zagreb,
Zagreb, Croatia
{stjepan.picek,domagoj.jakobovic}@fer.hr
[2] Radboud University Nijmegen, Nijmegen, The Netherlands
lejla@cs.ru.nl
[3] Riscure BV, Delft, The Netherlands
Buzing@riscure.com

Abstract. During recent years we observe an arms race between new creative methods for inserting effective faults and designing new countermeasures against such threats. Yet, even analyses of an unprotected smart card pose a problem for an analyst assuming constraints in time (or consequently, in a feasible number of measurements). In this paper we present a new kind of algorithm capable of finding faults in the black box test scenario - memetic algorithm. This algorithm combines the strengths of the following three algorithms: genetic algorithm, tabu search and local search. Furthermore, the same algorithm can be used if the goal is simply a rapid characterization of the search space. We compare our algorithm with random search and exhaustive search approaches. Experimental results show that our memetic algorithm is substantially more successful in both, locating faults and characterizing search space, than the other known methods. In reaching both goals, our memetic algorithm uses less than 300 measurements.

Keywords: Fault analysis · Glitches · Smart cards · Memetic algorithms

1 Introduction

Smart cards and other small pervasive devices such as RFID tags are used daily by billions of users for applications such as public transportation, Internet banking, online shopping, etc. The exposure to numerous threats, mainly coming from the adversary aiming at physical security, have led to this becoming one of the most actively researched topics by both academia and industry in the past two decades.

Anderson and Kuhn [1] put some doubt on the claimed tamper-resistance of smart cards almost two decades ago. This paper was shortly followed by the set of techniques for tampering with smart cards by Kömmerling and Kuhn [2].

In general, the techniques for tampering can be classified as *passive* or *active* [3]. In passive techniques some side-channel information is monitored

S. Mangard and A.Y. Poschmann (Eds.): COSADE 2015, LNCS 9064, pp. 159–173, 2015.
DOI: 10.1007/978-3-319-21476-4_11

while the card is supposed to work "normally". An example of these passive techniques is the analysis of power consumption, as introduced by Kocher et al. [4] or electromagnetic radiation [5]. In the case of active techniques, the device is not only monitored but also external interferences affect the normal behavior of the device. An example is Fault Injection (FI) attack. These interferences, the so-called *glitches*, can be of different nature: optical (laser pulses) and electrical glitches (voltage, clock), temperature changes, electromagnetic (EM) radiation, etc. They are used to cause malfunctioning, resulting in some cases in secret key recovery. Fault injection techniques by glitching are typically *non-invasive* techniques, in the sense that the smart card is not physically modified (in contrast to other *invasive* techniques that require hardware modifications).

A fault injection attack is considered to be successful if after exposing the device under attack to a specially crafted external interference, the device shows an unexpected behavior, which can be exploited by an attacker (e.g. leaking of sensitive information, bypassing security checks, etc.). However, this external insertion of signals has to be precisely tuned for the fault injection to succeed. As an example, a complete characterization of a clock signal glitch requires from the security analyst to define more than 10 parameters (related to clock signal voltage levels, time offset of the glitch, etc.,).

Finding the correct parameters for a successful FI can be considered as a search problem where one aims to find, within minimum time, the parameter configurations which result in a successful fault injection [6]. The search space, considering all possible combinations of the values of interest for the fault injection is typically too large to perform an exhaustive search.

Heuristic search algorithms can reduce the search time considerably. In this paper we investigate the feasibility of genetic algorithms (GAs) in this application domain. More specifically, we compare a standard GA with an enhanced GA called a memetic algorithm (MA), which adds local search iterations to the process. The motivation behind MA is the fact that GAs are generally good at *exploration* of the search space, but can often be improved in terms of the *exploitation* aspect.

Apart from introducing a GA framework to the FI domain we also make a distinction between the aim of finding as many glitches as possible on one hand, and characterizing the parameter space on the other (with characterization we consider identifying promising parameter regions). By finding these promising parameter regions and focusing the search on those values we increase the probability of the glitch candidates producing a successful glitch.

1.1   Related Work

The concept of fault analysis-based attacks is known in the research community for around twenty years. Boneh, DeMillo and Lipton published an attack on RSA where they exploit hardware faults for cryptanalysis [7,8]. Kömmerling and Kuhn present an extensive overview of techniques for fault injection and other tampering techniques and give ideas on how to mitigate some of them [2]. The paper highlights the case of power supply (VCC) fault injection (referred to as *glitch attacks*) and emphasizes those as the ones most useful in practice. Aumüller et al. performed one of the first practical works on fault analysis, in which they describe a real-life scenario of the impact of injecting glitches in the VCC and clock lines of an IC [9]. They also suggest some countermeasures applicable in this specific case. Approximately at the same time, Skoroboga-tov and Anderson introduce optical (laser) fault injection, where they describe injecting faults with a laser on a decapsulated IC [10]. This technique is still very successful nowadays for defeating the security of many protected devices, but it is out of scope for this work. Van Woudenberg et al. describe a real attack scenario for an Optical Fault Injection attack [11]. The practical problem of setting the parameters for fault injection is introduced in their work and the authors briefly discuss the lack of methodology to solve it as the main direction they rely on is based on heuristics. In addition, the paper gives a nice overview of all the practical issues that arise during a real execution of the FI attacks on actual hardware. Balasch et al. explore the effects of glitches injected in the clock line of an IC [12]. This work is very interesting for identifying various effects that a glitch can cause on real hardware in terms of defining all possible outcomes of a successful fault injection. However, it has to be noted that current smart cards usually run on an internal clock which makes this FI technique infeasible. The work of Boix Carpi et al. deals with a similar problem to ours but the authors take a different approach [6]. They use a self-adapting search algorithm that shows some potential when considering only two parameters, glitch shape and length. As a future direction they mention one could try genetic algorithms. Additionally, the same authors present preliminary work with genetic algorithms in [13]. As an extension to this work, we consider three parameters and take the analysis to the next level by unleashing the full power of evolutionary computation in combination with other search techniques.

1.2 Our Contribution

There are two main contributions in this paper. As far as we know, we are the first to use a hybrid (memetic) algorithm [14] to look for successful glitches. Our memetic algorithm combines techniques from genetic algorithms, local search and tabu search. The second contribution is that we use the same algorithm not only to find faults, but also for the characterization of the search space with a minimal number of measurements.

The remainder of this paper is organized as follows: in Sect. 2 we give our problem statement and relevant properties of the search space as well as smart card details. In Sect. 3 we present description of algorithms we use, and in Sect. 4 we give experimental results and a discussion. Finally, in Sect. 5 we offer conclusions and future work directions.

2 Preliminaries

In this section we start with a short introduction to the smart card used. Afterwards, we give information about possible verdict classes and search space parameters.

2.1 Smart Card Details

In all our experiments we use smart cards that are based on ATMega163+24C256 IC, realized in CMOS technology. Those cards do not deploy any side-channel or fault injection countermeasure. All processing on the card is performed in software, and cards are running on an external 1 MHz clock frequency.

For the experimental purposes, we attack a vulnerable PIN authentication mechanism. The PIN authentication mechanism is implemented as follows:

```
for (i = 0; i < 4; i++)
{
    if (pin[i] == input[i])
        ok_digits++;
}
if (ok_digits == 4)  //LOCATION FOR ATTACK
    respond_code(0x00, SW_NO_ERROR_msb, SW_NO_ERROR_lsb); //PIN IS CORRECT
else
    respond_code(0x00, 0x69, 0x85); //PIN IS WRONG
```

In the above code we want to glitch the target (smart card) while it is executing the second if-statement. However, we want to emphasize that our approach makes no assumptions on the software that runs on the smart card. First of all, we regard the target as a black box and only hypothesize that there exists a weakness in the implementation and that we can roughly estimate its location in time. Secondly, the most difficult part of finding good FI parameters are the electrical properties like glitch voltage and length. The right values for these two dimensions will make the target physically behave in an unspecified way, and most importantly, they will do so on any smart card of the same make (or production batch), regardless of the implemented software.

2.2 Verdict Classes and Boundaries

Fault injection testing equipment can output only verdict classes that correspond to successful measurements. There exist several possible classes for classifying a single measurement (i.e. attack attempt):

1. NORMAL: smart card behaves as expected and the glitch is ignored
2. RESET: smart card resets as a result of the glitch
3. MUTE: smart card stops all communication as a result of the glitch
4. INCONCLUSIVE: smart card responds in a way that cannot be classified in any other class
5. SUCCESS: smart card response is a specific, predetermined value that does not happen under normal operation

In the rest of this paper, we will consider RESET and MUTE classes as equivalent when interpreting the results. Additionally, when depicting graphs with measurement results, for each of classes we allocate a color. When the card responds NORMAL, we depict a green dot in the search space, for RESET/MUTE we depict a blue color, for INCONCLUSIVE yellow color and finally, for SUCCESS red color.

2.3   Search Space Parameters

There are multiple search space parameters that need to be set in the fault injection process. We informally divide those parameters into two groups. The first group consists of parameters that we influence with an external search space algorithm and are therefore of primary interest. The second group consists of parameters that we leave for fault injection framework to set randomly.

In the first group we include the following parameters: glitch length, glitch voltage and glitch offset. The glitch length refers to the time (ns) that the VCC line is perturbed. The glitch voltage is the number of miliVolts (mV) that is added to the VCC line. The glitch offset is the start time (ns) of the perturbation relative to the start of the clock cycle.

For example, suppose that the glitch length is 100 ns, the glitch voltage is -3 500 mV, the glitch offset is 250 ns, and the supplied VCC is 5 V. What will happen is that 250 ns after the start of the clock cycle the VCC line will be pulled down to 1.5 V for 100 ns, after which it will be restored to 5 V.

Those three parameters determine the *electrical* effect on the target: roughly speaking the product of glitch length and glitch voltage represents the amount of energy that is exerted on (or withheld from) the target. Since the current propagation within a clock cycle is not constant the offset timing is also important. We refer to the three parameters as the "shape" of the glitch. Note that this is a physical effect: if the glitch is too strong the target will "mute" or reset. On the other hand, if the glitch is too weak the target will not be disturbed, but if the glitch is of an appropriate shape the target will behave in an unspecified way. This is unrelated to the *logical* effect, which refers to the exact instruction that is being glitched.

The second parameter group covers the logical effect and it consists of the number of wait cycles and the number of glitch cycles. The wait cycles parameter refers to the number of clock cycles that are skipped before the glitch attack is performed, counting from the sending of the smart card command. The glitch cycles parameter specifies the number of successive clock cycles that are glitched.

For example, we could wait for 800 clock cycles and then apply the glitch in the next 5 cycles, assuming the relevant instruction is executed close to the foreseen time frame. The fact that the electrical effects can be observed independently of the logical effects allows the security analyst to work in two phases. First he will find a glitch "shape" that triggers the target to behave in an unspecified way. In the second phase he can search for the right wait cycles and glitch cycles values while applying the right glitch. In this paper we disregard the logical parameters. We assume reasonable ranges for the wait/glitch cycles and select uniformly at random from these ranges.

A useful property of the glitch shape parameters is that they display locality. The glitch offset has only a small range of values that "work". On top of that, the glitch voltage and the glitch length shows a monotonic behavior: once a glitch voltage is strong enough to force a card reset, then bigger values will also force a reset. The same goes for the glitch length. In practice this means that there is a clear phase transition in the voltage/length dimensions between NORMAL results and RESET/MUTE results. Also, the class of SUCCESS results (when offset is also guessed correctly) is often located around this phase transition. Note however that the exact effect of a glitch is stochastic and a perfect separation of parameter regions is impossible.

3 Approach and Methods

Looking for spots that lead to the successful fault injection can be considered as an optimization problem where we want to find as much "good" spots as possible in the minimal amount of time. The complexity of the problem depends on the number of considered parameters. However, before trying to give an answer about appropriate methods, we offer an illustration of the difficulty of the problem.

In a realistic setting that we consider, the glitch voltage parameter ranges from -5 000 mV to -50 mV, with a minimal step of 50 mV. The second parameter, glitch length, ranges from 2 ns to 150 ns with a step of 2 ns. If we conduct experiments where we are interested in only those two parameters and do exhaustive search, we need 7 400 measurements. If we assume that each measurement lasts one second, this gives us total time of two hours. However, if we add just one more parameter, e.g., glitch offset that goes from 100 ns to 400 ns with the 1 ns step, then we have in total 2.2 million measurements. This equals to more than 600 hrs of measurement time. Naturally, one also needs to take into account possible consequences for a smart card if it is tested for more than 600 hrs and the fact there are several more parameters of interest not even mentioned in this calculation.

The first objective targets at finding as many successful parameter combinations as possible, without regarding their values and their relation to each other. The second objective, on the other hand, aims to map the parameter space into regions with the same behavior outcome of the smart card. It is expected that the parameter combinations that result in the same behavior form regions in the search space which are adjacent to regions with different target behavior. Our experiments show that the region boundaries cannot be described with linear functions, and it is exactly along the boundaries that the successful attacks could be performed. Therefore, the objective of the search space characterization is to provide boundaries between different regions with as few measurements as possible.

When looking for faults we can expect that more attempts should be made with parameter values that resemble those that led to a fault, but from the other perspective, the analysis will use more measurements and will result in other regions less analyzed. On the other hand, when looking for a region of interest, we can expect that the algorithm will also find faults, but that behavior should not be specially rewarded.

3.1 Genetic Algorithm

Genetic Algorithms (GAs) belong to a subclass of evolutionary algorithms where the elements of the search space S are arrays of elementary types [15].

In our approach we need to change several parts of a standard GA in order to work with this specific problem setting. A 'standard' GA assigns fitness values to different points in the search space (individuals or potential solutions) and maintains a population of those, usually initialized randomly. A potential solution in this context represents the values of the three parameters in our search space. In each iteration (generation) it selects the better ones and eliminates the worse, combines different individuals to produce new ones (using the crossover operator) which replace the eliminated ones and randomly changes parts of new individuals (using mutation).

First, we need to map verdict classes to fitness values. Since the objective is to *maximize* the value of fitness function, we give higher values to verdict classes that are of a bigger importance. Observing that we are looking for parameters that behave differently from NORMAL behavior, for NORMAL class we give the smallest value of 1. RESET and MUTE classes we consider the same and we give them a value 2 since we expect to find faults in areas between NORMAL and RESET. Finally, for SUCCESS class we give a value 3. Since we are not able to define the INCONCLUSIVE class, we also assign it the same value as for the RESET/MUTE class.

Next, instead of a standard crossover operator we use the custom version - local crossover (LC). In this operator, the first crossover point (potential solution) is chosen randomly. The second point is chosen so the LC operator crosses two points that belong to *different* classes, and it generates a new offspring point between the parents. The position of the offspring point is chosen on the basis of the number of solutions in complete population that belong to the parents classes: a child is proportionally closer to the parent with the class that is *less represented*, i.e., the class with the smaller number of individuals in the population. Only in the case when the first parent belongs to SUCCESS class, this operator tries to find a second parent in the same class. A mutation is conducted by adding some random value to the parameters. We present a GA with aforementioned modifications as Algorithm 1.

3.2   Tabu Search

Since there exist only a few different verdict classes (and consequently, only a limited number of different fitness values) it is expected that a number of same solutions will emerge that may be tested repeatedly. Since each such solution leads to a unnecessary measurement, we adopt a technique from Tabu Search (TS) optimization method. Tabu Search works by declaring certain solution candidates that have already been visited as *tabu* and therefore not to be visited again [15]. The advantage of using the TS method is twofold in our case: first, we lower the total number of measurements performed and second, when not revisiting already visited locations, the algorithm is less likely to get stuck in a local optimum. We implement Tabu Search by using a list which stores all the solutions that have been already measured and allocated fitness values. If a new solution is created that is on the list, it is not measured but discarded immediately. Note that we do not implement all TS functionalities, but only those related with keeping the tabu list.

Algorithm 1 Genetic Algorithm.

Require: crx_count = 0, mut_count = 0
  repeat
    select first parent
    if first parent of SUCCESS class **then**
      try to find matching second parent
    else
      try to find second parent of different class
    end if
    perform crossover (depending on parent classes)
    copy child to new generation
    crx_count = crx_count + 1
  until $p_c * N \neq$ crx_count
  repeat
    select random individuals for tournament
    copy best of tournament to new generation
    mut_count = mut_count + 1
  until $(1 - p_c) * N \neq$ mut_count
  perform mutation on new generation with probability p_m
  evaluate population

3.3 Local Search

Local search (LS) is a metaheuristic method for solving computationally hard optimization problems [14]. Local search algorithms work on a single solution (instead of multiple solutions) and generally transcend only to neighbors of the current solution. It moves in the space of candidate solutions by applying local changes until it finds an optimal solution or the time bound is elapsed [14]. In our experiments we use one version of the divide-and-conquer algorithm where each new solution is located in the middle of parent solutions (binary search). In order to behave in such a manner, we need to define what is the space of candidate solution, i.e. in what neighborhood it can operate. To this end, we need to define an appropriate distance metric.

Two solutions are neighbors if they are at the distance smaller than d. In this paper we experiment with Euclidean [16] and Manhattan [17] distance metrics. Since the search space parameters are of different magnitudes, we use a normalized search range of [0,1].

Euclidean distance between two points a and b in an n-dimensional space is equals $d(a, b) = \sqrt{\sum_{i=1}^{n}(a_i - b_i)^2}$ [16].

Manhattan distance between two points is the sum of absolute differences of their Cartesian coordinates and it equals $d(a, b) = \sum_{i=1}^{n} |a_i - b_i|$. [17].

In our experiments we use a local search algorithm after each GA generation. The local search works on all pairs of individuals that are closer than the distance d and it runs while the distance between solutions is larger than the resolution r. With the resolution parameter we control how precise the characterization of the search space should be. We note that it was necessary to add two parameters to control the local search algorithm. The distance parameter d ensures that only individuals that are closer than d can participate in local search. The resolution parameter r controls when the LS should stop operating on each pair of individuals. Pseudocode for the LS is given in Algorithm 2.

Algorithm 2 Local Search algorithm.

---

create pool with all individuals ind
for all ind in the pool **do**
  select ind_tmp from the pool
  d = distance (ind, ind_tmp)
  if d > resolution and d < distance and class (ind) != class (ind_tmp) **then**
    make pair
    remove individuals from pool
  end if
end for
for all pairs **do**
  d = distance (ind_1, ind_2)
  if d > resolution **then**
    create point in between points
    call evaluator
    replace parent from the same class as offspring
  else
    remove pair
  end if
end for

3.4 Memetic Algorithm

Memetic Algorithms (MAs) represent a synergy between evolutionary algorithms (or any other population-based algorithm) and local improvement algorithms [15]. Most MAs can be interpreted as search strategies in which a population of solutions cooperate and compete [14].

In our experiments, the memetic algorithm is a combination of three aforementioned algorithms: genetic algorithm, tabu search and local search. Each of those algorithms should lend its strength to obtain a new, synergistic one that is more powerful than any of them individually. Genetic algorithms give their strength when finding promising regions in search space. Local search improves the convergence speed when looking for SUCCESS points (or regions between two verdict classes) and tabu search reduces the number of measurements by avoiding duplicate measurements.

4 Experiments and Results

In this section we present details about our experimental setup and the parameters considered. Afterwards, we present our results and give a short discussion. **Common parameters** for all experiments are given in Table 1.

Table 1. Common parameters.

Parameter	Parameter Value
Tournament size	3
Population size	30
Stopping criterion	10 generations
Mutation rate	0.1
Glitch length	[2, 150] ns
Glitch voltage	[-5 000, -50] mV
Glitch offset	[100, 400] ns
Glitch cycles	random from [1, 10]
Wait cycles	random from [750, 850]

As it can be observed in Table 1 we use a small number of generations and a small population size since we are interested in a rapid characterization or finding faults. Indeed, if one has sufficient time at his disposal no method can outperform exhaustive search.

4.1 Experimental Results

When conducting experiments, we compare our results with random search and exhaustive search methods. Here we give the results for the two methods.

Random Search. In this method search space parameters are chosen uniformly at random. Figure 1(a) displays random search with 2 500 measurements.

Exhaustive Search. In order to check the full characterization of search space we also run an exhaustive search algorithm. Here, parameters of interest are glitch voltage, length and offset. Since there are too many possible solutions for any realistic exhaustive search we conduct exhaustive search for glitch length and voltage while other parameters are chosen uniformly at random. Figure 1(b) shows the results of 7500 measurements.

Next, we present results of our new algorithms separately for the case where the goal is to find as many faults as possible and for the case where the goal is the characterization of search space. After a short tuning phase we set distance d parameter to the value of 0.3 and resolution r parameter to the value of 0.1 since with those values we observe the best behavior. However, our experiments also show that these parameters are quite robust and small changes in values do not significantly change algorithm performance.

4.2 Finding Faults

When the goal is finding faults, we conduct several runs of different algorithm versions and then we present averaged values. Columns Normal, Reset and

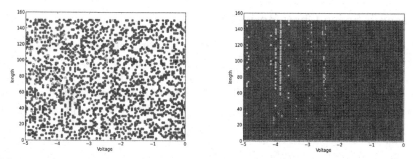

(a) Random search, 2 500 measurements (b) Exhaustive search, 7 500 measurements

Fig. 1. Measurements for random and exhaustive search methods.

Success show average number of NORMAL, RESET/MUTE and SUCCESS measurements. In Table 2 we give the results for three different versions of our algorithm where we can see that GA+TS+LS algorithm with Euclidean distance metric finds the most SUCCESS points on average.

In Figs. 2(a) and 2(b) we give an example of one run of GA+TS+LS algorithm with Euclidean distance and 250 measurements. In this experiment, with 250 measurements in total, we found 21 glitches which represents 8.5 % of total measured points.

Table 2. Average results of experiments.

Algorithm	Normal (%)	Reset (%)	Success (%)
GA+TS	58.08	39.97	1.94
GA+TS+LS, Euclidean	55.29	41.87	2.84
GA+TS+LS, Manhattan	62.76	36.45	0.78

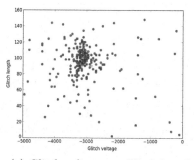

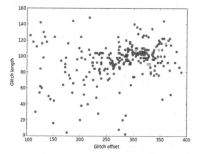

(a) Glitch voltage vs. Glitch length   (b) Glitch offset vs. Glitch length

Fig. 2. GA+TS+LS, 250 measurements.

4.3   Search Space Characterization

When the goal is to characterize the search space, or more precisely the region between NORMAL and RESET/MUTE classes, we are not interested in SUC-CESS points. Therefore, we can treat them as NORMAL or RESET/MUTE points (and give them fitness values 1 or 2, respectively). Again in this case, the number of measurements is set to 250. In Figs. 3(a) to 3(d) we present results for search space characterization with four different algorithms.

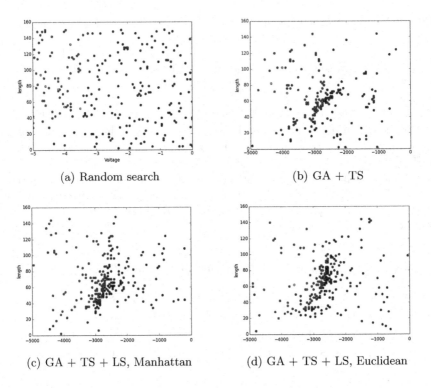

(a) Random search

(b) GA + TS

(c) GA + TS + LS, Manhattan

(d) GA + TS + LS, Euclidean

Fig. 3. Algorithms for the space characterization, 250 measurements.

We see that random search is not capable to characterize interesting regions with a small number of measurements. A combination of GA, TS and LS algorithms with Manhattan distance performs best since it accurately describes the longest part of the interesting region.

When observing differences in regards to distance metrics, we see that Euclidean distance gives better results when finding faults while Manhattan distance is better in space characterization scenario. However, this observation should be considered *cum grano salis* since we use the same distance value in both cases. It can be concluded that the smaller distances are better when looking for specific points (smaller distances are to be expected in Euclidean metric due to the squaring operation of normalized values) while bigger distance values can cover more space and characterize it better as it can be seen from the case where we use Manhattan distance. As evident from the results, the memetic algorithm behaves much better than the genetic algorithm considered (although, the GA we use is specialized and already behaves much better than the standard GA).

It is very difficult to give a meaningful comparison between the efficiency of our algorithm and for instance algorithms presented in [6,13] due to several reasons. First, we add one more dimension (glitch offset) to the search space and thus we render some of the operators from previous works non applicable. Moreover, our problem is much more difficult due to the extra dimension. Although exhaustive search in two dimensions (with some parameter steps) would take several hours we still consider it to be a realistic approach while with three parameter dimensions this problem becomes completely non practical in a realistic environment. With the increase of the number of parameters, the methods presented here need no additional adjustment and should prove even more efficient with regard to random search, which remains to be addressed.

Next, in our approach we set strict constraints on the available number of measurements which was not the case in previous works (there the goal was a minimal number of measurements without explicitly stating the minimal number). Furthermore, as GAs use information from known solutions in future generations, after finding several faults we can expect to find asymptotically more faults in future generations and that probability increases with the number of generations. In related work there is also a distinction that they conduct three measurements per point to check for CHANGING class [6]. In our approach we do not consider CHANGING class and consequently we do not conduct multiple measurements of the same points.

Lastly, based on the results in [6] it seems that some of the SUCCESS points that are found and taken into account in statistics are actually repeated measurements of the same points. Since in our approach Tabu Search renders that impossible, it would not be possible to compare those results without removing TS constraint from our algorithm. As evident from our results, TS on average reduces the total number of measurements by more than 20 % which results in more unique points our algorithm can generate.

5 Conclusions and Future Work

In this work we revisit the problem of fiddling with multiple parameters for successful fault injection. Our experiments with the memetic algorithm show that one can successfully find faults with a limited number of measurements. Additionally, our algorithm can be used to characterize interesting search space regions. Both scenarios are explored with a "small" i.e. feasible number of measurements. By adding more measurements (and therefore GA generations) we obtain even better results since the GA works by using existing solutions to find new, better solutions. We do not claim that the GA (or the memetic algorithm) is the best possible method, but we demonstrate there are nature-inspired algorithms that can significantly improve the FI process.

Glitch testing in this work has only been performed on a target with no countermeasures. Since it is expected that the search space is affected by such countermeasures, e.g. glitch sensors, the applicability of this approach in a real world attack scenario remains to be assessed. A possible step in our research is therefore to experiment with smart cards on which countermeasures against FI are implemented.

Acknowledgments. This work was supported in part by the Technology Foundation STW (project 12624 - SIDES), The Netherlands Organization for Scientific Research NWO (project ProFIL 628.001.007) and the ICT COST action IC1204 TRUDEVICE.

References

1. Anderson, R., Kuhn, M.: Tamper resistance – a cautionary note. In: Proceedings of the Second Usenix Workshop on Electronic Commerce, pp. 1–11 (1996)
2. Kömmerling, O., Kuhn, M.G.: Design principles for tamper-resistant smartcard processors. In: Proceedings of the USENIX Workshop on Smartcard Technology on USENIX Workshop on Smartcard Technology, ser. WOST 1999. Berkeley, CA, USA: USENIX Association, p. 2 (1999)
3. Mangard, S., Oswald, E., Popp, T.: Power Analysis Attacks. Revealing the Secrets of Smart Cards (Advances in Information Security). Springer-Verlag New York Inc., Secaucus (2007)
4. Kocher, P.C., Jaffe, J., Jun, B.: Differential power analysis. In: Wiener, M. (ed.) CRYPTO 1999. LNCS, vol. 1666, p. 388. Springer, Heidelberg (1999)
5. Quisquater, J.-J., Samyde, D.: ElectroMagnetic Analysis (EMA): measures and counter-measures for smart cards. In: Attali, S., Jensen, T. (eds.) E-smart 2001. LNCS, vol. 2140, p. 200. Springer, Heidelberg (2001)
6. Carpi, R.B., Picek, S., Batina, L., Menarini, F., Jakobovic, D., Golub, M.: Glitch it if you can: parameter search strategies for successful fault injection. In: Francillon, A., Rohatgi, P. (eds.) CARDIS 2013. LNCS, vol. 8419, pp. 236–252. Springer, Heidelberg (2014)
7. Boneh, D., DeMillo, R., Lipton, R.: New threat model breaks crypto codes. Bellcore 85 Press Release (1996)
8. Boneh, D., DeMillo, R.A., Lipton, R.J.: On the importance of checking cryptographic protocols for faults. In: Fumy, W. (ed.) EUROCRYPT 1997. LNCS, vol. 1233, pp. 37–51. Springer, Heidelberg (1997)
9. Aumüller, C., Bier, P., Fischer, W., Hofreiter, P., Seifert, J.-P.: Fault attacks on RSA with CRT: concrete results and practical countermeasures. In: Kaliski, B.S., Koç, K., Paar, C. (eds.) CHES 2002. LNCS, vol. 2523, pp. 260–275. Springer, Heidelberg (2003)
10. Skorobogatov, S.P., Anderson, R.J.: Optical fault induction attacks. CHES 2002. LNCS, vol. 2523, pp. 2–12. Springer, Heidelberg (2003)
11. van Woudenberg, J., Witteman, M., Menarini, F.: Practical optical fault injection on secure microcontrollers. In: 2011 Workshop on Fault Diagnosis and Tolerance in Cryptography (FDTC), pp. 91–99 (2011)
12. Balasch, J., Gierlichs, B., Verbauwhede, I.: An in-depth and black-box characterization of the effects of clock glitches on 8-bit MCUs. In: Proceedings of the 2011 Workshop on Fault Diagnosis and Tolerance in Cryptography, ser. FDTC 2011. IEEE Computer Society, Washington, DC, USA, pp. 105–114 (2011)

13. Picek, S., Batina, L., Jakobovic, D., Carpi, R.B.: Evolving genetic algorithms for fault injection attacks. In: 2014 Proceedings of the 35th International Convention, MIPRO 2014, Opatija, Croatia, 26–30 May 2014. IEEE (2014)
14. Glover, F.W., Kochenberger, G.A. (eds.): Handbook of Metaheuristics. International Series in Operations Research & Management Science, vol. 114, 1st edn. Springer, Heidelberg (2003)
15. Weise, T.: Global Optimization Algorithms - Theory and Application, 2nd ed. (2009). http://www.it-weise.de/
16. Fabbri, R., Costa, L.D.F., Torelli, J.C., Bruno, O.M.: 2d euclidean distance transform algorithms: a comparative survey. ACM Comput. Surv. 40(1), 2:1–2:44 (2008)
17. Krause, E.F.: Taxicab Geometry: An Adventure in Non-Euclidean Geometry. Dover Books on Mathematics. Dover Publications, New York (1988)

Differential Fault Intensity Analysis on PRESENT and LED Block Ciphers

Nahid Farhady Ghalaty[✉], Bilgiday Yuce, and Patrick Schaumont

Bradley Department of Electrical and Computer Engineering,
Virginia Polytechnic Institute and State University,
Blacksburg, VA, USA
{farhady,bilgiday,schaum}@vt.edu

Abstract. Differential Fault Intensity Analysis (DFIA) is a recently introduced fault analysis technique. This technique is based on the observation that faults are biased and thus are non-uniformly distributed over the cipher state variables. The adversary uses the fault bias as a source of leakage by controlling the intensity of fault injection. DFIA exploits statistical analysis to correlate the secret key to the biased fault behavior. In this work, we show a DFIA attack on two lightweight block ciphers: PRESENT and LED. For each algorithm, our research analyzes the efficiency of DFIA on a round-serial implementation and on a nibble-serial implementation. We show that all algorithms and all implementation variants can be broken with 10 to 36 fault intensity levels, depending on the case. We also analyze the factors that affect the convergence of DFIA. We show that there is a trade-off between the number of required plaintexts, and the resolution of the fault-injection equipment. Thus, an adversary with lower-quality fault-injection equipment may still be as effective as an adversary with high-quality fault-injection equipment, simply by using additional encryptions. This confirms that DFIA is effective against a range of algorithms using a range of fault injection techniques.

Keywords: Differential attack · Fault intensity · Light-weight block cipher · PRESENT · LED

1 Introduction

Nowadays, lightweight cryptographic primitives are recommended to be used to secure various resource-constrained systems such as RFID tags and sensor networks [1,2]. The security of a cryptographic primitive relies on both its algorithmic features and on its physical implementation.

Physical attacks are divided into two groups. Side Channel Attacks retrieve the secret key by using statistical tests on the information leaked from the cryptographic device during its execution [3]. Fault attacks, first, intentionally disturb a cryptographic device by means of fault injection to induce errors in the output of the device. Then, they exploit the erroneous outputs to mathematically reverse-engineer the secret key [4].

© Springer International Publishing Switzerland 2015
S. Mangard and A.Y. Poschmann (Eds.): COSADE 2015, LNCS 9064, pp. 174–188, 2015.
DOI: 10.1007/978-3-319-21476-4_12

Differential Fault Intensity Analysis (DFIA) is a recently introduced fault analysis technique [5]. In DFIA, the attacker injects faults by means of intentional variation of the fault intensity. Using this fault injection technique, he induces biased faults in the intermediate state of the cryptographic algorithm. Under the biased fault model, a gradual change in the fault intensity will cause a small change in the faulty state variable. Using the faulty ciphertext, the attacker computes the key-dependent secret state variable under each key hypothesis. Finally, he performs statistical tests on each of the computed state variables and selects the key guess that is most likely under the biased fault model. Due to the non-linear transformations of the cipher, the correct key hypothesis shows only small changes on the variable, while the wrong key guesses show a random behavior. This attack combines the principles of Differential Power Analysis and fault injection.

In this paper, we demonstrate a DFIA attack on two lightweight cryptographic algorithms: PRESENT [6] and LED [7]. In contrast to AES, PRESENT and LED are nibble-oriented (4-bit). This makes the observation and exploitation of biased faults more difficult. We therefore investigate the feasibility of DFIA on both nibble-serial and round-serial implementations of PRESENT and LED. We evaluate the practicality of the biased fault model and the attack strategies on both nibble-serial and round-serial implementations of the algorithms.

Our results show that a single plaintext and 10 fault intensity levels are sufficient to extract the key of a nibble-serial PRESENT-80 design. We also show that 12 fault intensity levels are sufficient to extract the key of a round-serial PRESENT-80 design. Besides a DFIA on PRESENT-80, the paper also provides the attack results for PRESENT-128, LED-80, and LED-128. We confirm that all these designs can be broken.

We also demonstrate that DFIA [5] can be easily extended over multiple plaintexts, and that this increases the efficiency of the attack in narrowing down the key search space. We show that using multiple plaintexts can compensate for the low-resolution fault injection equipments. Therefore, DFIA can still retrieve the correct key efficiently, even if the attacker is not in possession of a high-quality fault injection tool.

The paper is organized as follows. Section 2 describes the DFIA and its fault model requirements. In this section, we also explain the PRESENT and LED algorithms and the nibble-serial and round-serial implementations of these algorithms. Section 3 explains the DFIA attack procedure on the PRESENT and LED algorithms. Section 4 shows the required number of fault injections for a DFIA attack on the PRESENT and LED. In this section, we also show the efficiency of the extended version of the DFIA. Section 5 covers the previous work that relies on fault bias. Section 6 concludes the paper.

2   Background and Notation

This section explains the principles of the DFIA method. We will first explain the concept of biased fault and an easy way to control it.

Table 1. Symbols of DFIA attack procedure

P	Plaintext
Q	Total number of injected faults with different intensities
q	A specific fault intensity
C'_q	Faulty ciphertext under fault intensity q
S	Correct state
k	Key hypothesis
$S'_{k,q,P}$	Faulty state under hypothesis $K = k$, fault q, $S'_{k,q,P} = f(C'_q, k)$ and input P

2.1 Fault Model

The fault model is a combination of three factors. These factors are fault location, fault timing and fault type. Fault location and fault timing define the spatial and temporal location of the fault in a hardware circuit, respectively. The fault type describes the behavior of the injected fault, and can be stuck-at, set-reset, random bit-flip, or biased fault respectively. Throughout this paper, we refer to following terms and definitions:

- Fault Intensity: Fault intensity is the strength by which a circuit is pushed outside of its nominal operating conditions with the intent of inducing a fault. For example, when faults are introduced using clock glitches, then the fault intensity corresponds to the shortened clock cycle that is obtained as a result of the glitches.
- Fault Sensitivity: The fault sensitivity is the fault intensity at which a hardware circuit reflects faulty behavior [8]. For example, when faults are injected by means of clock glitches, then fault sensitivity generally corresponds to the critical path of the circuit.
- Biased Fault: A biased fault is the incremental fault behavior obtained as a result of gradual increase in fault intensity. For DFIA, we are especially interested in using minimal fault bias (e.g. changes of one or two bits in a state variable), although other authors have shown that any fault bias is a source of leakage [9].

One of the cheapest and most convenient methods of injecting biased faults into a hardware device is clock glitching. In this method, the attacker creates biased faults via injecting glitches into the clock signal of the device. To make a circuit fail its timing constraints,the attacker gradually increases the fault intensity by decreasing the clock period via glitch injection. As a result, he can obtain biased faults because of the existing non-uniformity in the path delays of the circuit.

2.2 Differential Fault Intensity Analysis Using Multiple Plaintexts

This section summarizes Differential Fault Intensity Analysis. Algorithm 1 describes the attack procedure, and Table 1 lists the symbols used in this paper.

Algorithm 1. DFIA Attack Procedure using Multiple Plaintext

Assume *Cryptographic Algorithm, Fault Injection Tool*;
Result *Correct Key Guess* ;
foreach *Plaintext P* **do**
  foreach *Faultintensity q, $1 \leq q \leq Q$* **do**
    Obtain faulty ciphertext C'_q;
    foreach *Key Hypothesis k* **do**
      Compute faulty state hypothesis $S'_{k,q,P} = f(C'_q, k)$;

//Post-processing phase ;
foreach *Key Hypothesis k* **do**
  Calculate $\rho_k = \sum_P \sum_{n=1}^{Q} \sum_{m=1}^{n-1} HD(S'_{k,n,P}, S'_{k,m,P})$;
$K = \min \rho_k$;

DFIA starts by applying a fault intensity q into an intermediate value S. The attacker next observes the faulty ciphertext C'_q, and derives the faulty intermediate value $S'_{k,q,P} = f(C'_q, k)$ under a key hypothesis k. The attacker repeats these two steps for Q different fault intensities by gradually increasing the fault intensity each time. In the post-processing step, for each key hypothesis, he computes the cumulative Hamming Distance among all faulty intermediate values. Finally, the attacker selects the key hypothesis that corresponds to the minimum cumulative Hamming Distance. The reason of looking for minimum is that for the correct key hypothesis, the cumulative Hamming Distance is correlated with the fault intensity, and thus, it is minimal. A wrong key hypothesis infers a larger, random cumulative Hamming Distance due to the non-linear diffusion and confusion properties of the attacked cipher. Hence, the correct key results in the minimum cumulative Hamming Distance as long as the applied fault intensities induce biased faults. Ghalaty et al. [5] show this behavior on AES for different biased fault injection scenarios. They draw two conclusions. First, DFIA converges for any given set of biased faults. Second, DFIA converges faster for strongly-biased faults (e.g. 1-bit faults) than it does for weakly-biased faults (e.g. 4-bit faults).

The original DFIA is applied using a single plaintext value [5]. However, DFIA can be easily extended to multiple plaintexts, by repeating the above steps for each plaintext, and by accumulating the resulting Hamming Distance values for each key hypothesis. Again, the global minimum will be obtained only under the correct key hypothesis. In this paper, we make use of this feature, and we show that it can be used to improve the efficiency of DFIA when few biased faults are available, or when the fault injection equipment has limited resolution.

2.3 PRESENT Block Cipher

We make a brief overview of PRESENT and our implementations of it. PRESENT is a lightweight block cipher that was recently standardized by

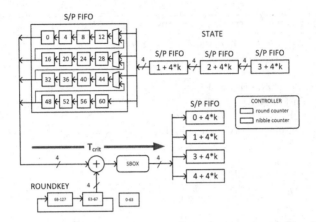

Fig. 1. Nibble-serial implementation of PRESENT

IEEE [6]. It uses an SP-network structure, and loosely follows the structure of AES, with the following important differences. It has 31 rounds, and uses a block-length of 64 bits. It uses a selectable key size of 80 bit or 128 bit, and both versions are distinguished through their name (PRESENT-80 or PRESENT-128). Each round consists of three steps, including a roundkey addition layer, a nonlinear substitution layer with sixteen 4-bit Sbox, and a permutation layer. After the last round, an additional post-whitening step is included by adding a final roundkey.

The 64-bit roundkey is extracted from the upper part of the key register, and each round the key is updated with a key-size dependent key scheduling algorithm. The key schedule for PRESENT-80 is shown in Eqs. (1a) through (1c). The key schedule for PRESENT-128 is slightly more complex, and can be consulted in [6].

$$K_{79}K_{78}....K_0 = K_{18}K_{17}....K_{19} \tag{1a}$$

$$K_{79}K_{78}K_{77}K_{76} = Sbox[K_{79}K_{78}K_{77}K_{76}] \tag{1b}$$

$$K_{19}K_{18}K_{17}K_{16}K_{15} = K_{19}K_{18}K_{17}K_{16}K_{15} \oplus round\_counter \tag{1c}$$

In this work, we studied both a round-serial and a nibble-serial implementation. The reason for this is to show the feasibility of DFIA on different implementations of the same cipher. The round-serial implementation computes an entire round of a complete block in a single clock cycle. This implementation is straightforward and follows the design of the original PRESENT paper [6]. We also developed a nibble-serial design, as shown in Fig. 1. In this case, one round for a single nibble (4 bits) from a block is computed in a single clock cycle, and this requires sequentialization of the round operations. This is easy to achieve for the roundkey addition and the Sbox substitution. For the permutation layer, we make use of the property that PRESENT's permutation is a 4-bit by 16-bit transpose operation: 4 bits of the permutation output are taken

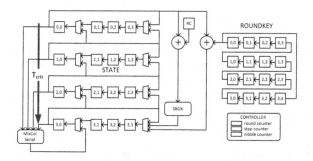

Fig. 2. Nibble-serial implementation of LED

from a column of 4-bits of an input block, when the block is arranged as a 4-bit by 16-bit matrix. In Fig. 1, we implement the permutation using serial/parallel FIFO modules, which consist of four 4-bit FIFO's that either operate as a single 16-bit FIFO (serial mode) or else as four parallel 4-bit FIFO's (parallel mode). A complete block is stored in four serial/parallel FIFOs. Using two such structures, which store either the odd or even round states, a compact nibble-serial version of PRESENT is obtained.

Of particular note for our fault analysis is the critical path in these structures. The critical path runs through the Sbox and roundkey addition operations. For the round-serial design, all Sbox operations will be in the critical path in a given clock cycle. For the nibble-serial design, on the other hand, only a single Sbox operation will be in the critical path in a given clock cycle.

2.4   LED Block Cipher

The Light Encryption Device (LED) is a compact block cipher that was developed after PRESENT, and that integrates further insight into the lightweight cipher design process [7]. This block cipher, too, is an SPN structure, with a 64-bit block size. It supports two different key sizes, 64-bit or 128-bit, and the notation LED-64 and LED-128 is used to distinguish these cases. LED-64 has 8 steps of 4 rounds each, for a total of 32 rounds. In between steps, roundkeys are added. Each of the rounds includes operations similar to AES (AddConstants(AC), SubCells(Sbox), ShiftRows(SR), MixColumnSerial(MC)), but each of these steps is specifically optimized towards lightweight encryption. LED organizes the state as a four by four matrix of nibbles, and the round operations operate on these nibbles.

The LED cipher does not use a key scheduling algorithm. Rather, it reuses the same key for every step. In the case of 128-bit key, the key bits are divided into two groups and each round uses one of them alternatively. LED includes a post-whitening step with a final addroundkey.

As with PRESENT, we developed a round-serial and a nibble serial version of LED for DFIA analysis. Figure 2 shows the architecture of the nibble-serial design. It follows the design guidelines of the original LED paper [7]. The State is organized in a FIFO-like structure of 16 nibbles. The structure can rotate the first column to compute MixColumnSerial, and it can rotate rows to compute ShiftRows. SubCells and AddRoundKey rotate the entire matrix through an Sbox and round-key addition respectively. The critical path runs through the MixColumnSerial. This is true for either the nibble-serial as well as the round-serial design. Fault injection using glitches will directly affect the variables computed in the critical path.

2.5   Implementations of the Block Ciphers

We wrote Verilog codes for our block cipher designs, namely, round-serial LED (LED-rs), nibble-serial LED (LED-ns), round-serial PRESENT (PRE-rs), and nibble-serial PRESENT (PRE-ns). We choose the key size as 128-bit in our implementations. We also generated gate-level netlist files for an Altera Cyclone IV FPGA (60 nm Technology). We use these netlists for gate-level simulations, which are carried out using Modelsim-Altera 10.1d [10] software, to verify our claims throughout the paper.

3   DFIA Attack on PRESENT and LED

In this section, we explain the DFIA attack on nibble-serial and round-serial implementations of PRESENT and LED block ciphers. To get the full key, the attacker must perform DFIA for the last two rounds of PRESENT-80 (i.e. round 30 and 31) and the last three rounds of PRESENT-128 [11]. The LED cipher has a very simple key scheduling method, and thus, we can retrieve the key by attacking the last round of LED-64. For LED-128, we have to attack the last two rounds to retrieve the key.

DFIA has two phases: Injecting biased faults into the intermediate state of the block cipher and post-processing the faulty ciphertexts to retrieve the key. The biased fault injection is nibble-wise (i.e., 4-bit) for nibble-serial implementations, while it is state-wise (i.e., 64-bit) for round-serial implementations. Regardless of DFIA on round-serial or nibble-serial designs, the post-processing is always applied on a single key nibble at a time.

3.1   Biased Fault Injection in PRESENT and LED

The proposed DFIA attacks build upon injecting biased faults in the inputs of Sbox blocks. One can use a clock glitch injection method such as in Fig. 3(a) for this purpose. This method generates an input clock signal for the circuit as a combination of two clock signals, namely, glitch clock (clk_g) and nominal clock (clk_o). As it is seen in Fig. 3(b), we inject glitches in the clk_o via an enable signal (g_en). To inject a biased fault in the input of an Sbox, we set the g_en signal just before the clock cycle, in which the Sbox is employed. Such a glitch injection makes some timing paths fail and causes a biased fault in the input of

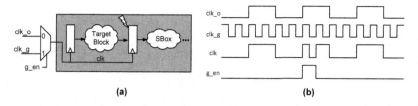

Fig. 3. (a) Block diagram of experimental setup (b) Timing diagram of experimental setup

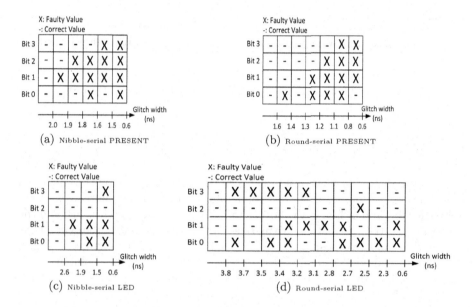

Fig. 4. Biased fault in the PRESENT and LED implementations

the Sbox. We control the fault intensity by increasing/decreasing the frequency of the *clk_g* signal.

The target block of DFIA is different for each implementation. For LED-ns, the target block is MixColumnSerial (MC) logic. We create biased faults in the outputs of the MC logic by violating its timing paths. Then, the biased faults are transferred to the inputs of Sbox blocks via linear AddConstants (AC) layer. The target block of PRE-ns is the roundkey addition and substitution blocks. For LED-rs and PRE-rs the target blocks are the whole round logic of the corresponding algorithms.

3.2   Biased Faults in PRESENT and LED Exist

In this section, we present a set of experimental results to verify that fault bias is a feasible fault source. We demonstrated biased faults through gate-level

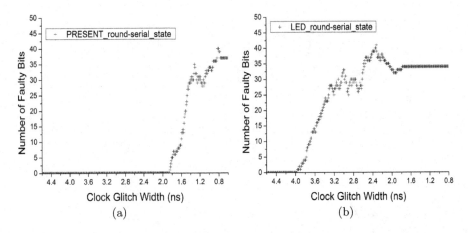

Fig. 5. Biased fault injection on state of (a) PRESENT and (b) LED

(post-place-and-route) simulation of the four block cipher implementations. In Fig. 4, we present the results for Sbox of the four implementations for a single plaintext.

Figure 4 shows the relationship between the clock glitch width and obtained faults in the Sbox inputs at the last round of the corresponding implementation. For each subgraph of Fig. 4, the horizontal axis is the clock glitch width and the vertical axis is the bit position. We mark a faulty bit position with the symbol (X) and mark a fault-free position with the symbol (-). In each subgraph of Fig. 4, we observe a minimal Hamming Distance between two neighbor columns. This behavior verifies the existence of fault bias in our implementations.

In Figs. 5(a) and 5(b), we show the number of faulty bits that are induced in the 64-bit state with respect to the clock glitch width for PRE-rs and LED-rs, respectively. These two graphs show that the fault bias exists for the 64-bit state as well. The PRE-rs will fail at higher fault intensity (i.e. at a narrower glitch width) than LED-rs. The reason is that the critical path of PRE-rs is shorter compared to the LED-rs. Thus, the attacker needs higher-capability fault injection tool to inject fault into PRE-rs.

3.3 Post-processing of DFIA on PRESENT

In this section, we describe the procedure to retrieve the key for PRE-ns and PRE-rs implementations. To obtain the 80-bit key of PRESENT-80, we first retrieve the round key of round 31 to get the 64 most significant bits of the key. Then, to retrieve the remaining key bits, we retrieve the round key of round 30. Similarly, for PRESENT-128, the attacker must retrieve the round keys of rounds 31, 30, and 29.

We can retrieve each nibble of a round key separately. Therefore, the key retrieval procedure for nibble-serial and round-serial implementations is the same. Following is the procedure to retrieve the 80-bit key for PRESENT-80.

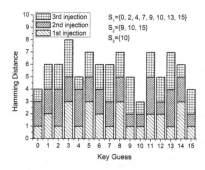

Fig. 6. DFIA steps to retrieve 4-bit key of PRESENT

We assume that the attacker has already collected the required amount of faulty ciphertexts to retrieve the key (using the method in Sect. 3.1).

The DFIA attack on PRESENT-80 follows Algorithm 1 as explained earlier. The faulty state variable is computed using Eq. 2. By repeating this process for all nibbles of round 31, the attacker can retrieve the correct value for $K_{79}K_{78}....K_{16}$. In order to retrieve $K_{16}K_{15}....K_0$, the attacker has to process 4 least significant nibbles of round 30 as well.

$$S'_{k,C'} = PlayerInv(SboxInv(C' \oplus K)) \tag{2}$$

Figure 6 shows an example DFIA attack to guess a nibble of a key. In this figure, the attacker injects four fault intensities: no injection, 1-bit fault injection, 2-bit fault injection and 3-bit fault injection. In this example, we retrieve one nibble of the round key with three fault injections. The bottom section of the bar chart shows the Hamming Distance between the first two intensities. The candidates for the correct key guess are the key guesses that show the minimum Hamming Distance, which is the set $G_1 = \{0, 2, 4, 7, 9, 10, 13, 15\}$ after the 1-bit fault injection. The middle section of the bar chart shows the Hamming Distance between 1-bit fault injection and 2-bit fault injection. As it is seen, the set of key candidates for correct key guess reduces to the set $G_2 = \{9, 10, 15\}$ after the 2-bit fault injection. The top section of the bar chart shows the Hamming Distance between 2-bit fault injection and 3-bit fault injection. The last fault injection gives us the unique key guess, which is $G_3 = \{10\}$.

3.4 Post-processing of DFIA on LED

In this section, we describe the procedure to retrieve the key for LED algorithm for nibble-serial and round-serial implementations. As the LED uses a very simple key scheduling method, the key can be retrieved by attacking the last round of LED-64. For LED-128, the attacker can retrieve the most significant 64-bit of the key by attacking the round 31. Then, he can retrieve the remaining bits by attacking the round 30.

The post-processing step of LED is different than the post-processing of PRESENT because LED includes a MixColumnSerial operation in its last round. The MixColumnSerial operation spreads the single faulty nibble in the intermediate state (S) to four nibbles of the faulty ciphertext (C'). Therefore, the reconstruction of the hypothesized faulty intermediate state now requires a hypothesis on 16 key bits, which means that we have 2^{16} different hypotheses. We can solve this problem via a method proposed by Jeong et al. [12]. The solution relies on peeling off the MixColumnSerial operation of the last round by using an equivalent ciphertext (C'^*), and retrieving an equivalent key (K^*) instead of the actual key (K). The equivalent key and ciphertext satisfy the Eqs. 3a and 3b, respectively.

$$K^* = MCInv(K) \tag{3a}$$

$$C'^* = MCInv(C') \tag{3b}$$

As Eq. 3b removes the effect of MixColumnSerial operation on C', one faulty nibble in S corresponds to one faulty nibble in C'^*. Therefore, we can use the C'^* to retrieve each nibble of the K^* with 4-bit key hypotheses. Using S' and C'^*, we can perform DFIA (Algorithm 1) to retrieve four bits of the K^* using Eq. 4. By repeating the procedure for 16 nibbles of the C'^*, we retrieve all 16 nibbles of the K^*. Then, we apply the MixColumnSerial operation on the K^* to retrieve the actual key K.

$$S' = SboxInv(SRInv(C'^* \oplus K'^*)) \tag{4}$$

The validity of the described solution can be seen from Eqs. 5a through 5d. The faulty ciphertext C' is computed by Eq. 5a. Equation 5b is obtained by applying the MixColumnsInverse operation to both sides of Eq. 5a. Using the distributive property of the MixColumnsSerial over the XOR operation, we obtain Eq. 5b. Using Eqs. 3b and 3a we obtain Eq. 5d.

$$C' = MC(SR(Sbox(S'))) \oplus K \tag{5a}$$

$$MCInv(C') = MCInv(MC(SR(Sbox(S'))) \oplus K) \tag{5b}$$

$$MCInv(C') = MCInv(MC(SR(Sbox(S')))) \oplus MCInv(K) \tag{5c}$$

$$C'^* = SR(Sbox(S')) \oplus K^* \tag{5d}$$

4   Results

We evaluated the proposed DFIA attacks using gate-level simulation. In our gate-level simulations, we first generated 50 random plaintexts. Then, for each of the four implementations, we obtained the ciphertexts for different clock glitch widths. In this experiment, we gradually decreased the clock glitch width from 4.6 ns to 0.6 ns with 100 ps step size. At the end, we obtained 40 ciphertexts for

Table 2. Required Number of Physical Fault Intensity Levels and Glitched Clock Cycles for DFIA Attack on PRESENT and LED with 100 ps Fault Injection Resolution

	# of Fault intensity levels		# of Glitched clock cycles	
	Nibble-serial	Round-serial	Nibble-serial	Round-serial
PRESENT-80	10	12	160	12
PRESENT-128	16	18	256	18
LED-64	14	18	224	18
LED-128	28	36	448	36

each plaintext and each implementation. As it can be seen from previous work, the selected step size is a reasonable value [13]. We present the analysis of our results in the following subsections. We also study the trade-off between glitch resolution and using multiple plaintexts in DFIA.

4.1   Results of DFIA on PRESENT and LED

Table 2 shows the results of a DFIA attack on PRESENT and LED implementations. The first two columns in Table 2 show the maximum number of required fault intensity levels for each implementation to get enough faulty ciphertexts in each nibble. For example, in nibble-serial PRESENT-80, the attacker is required to increase the fault intensity 10 times to get enough faulty ciphertexts in each nibble. The obtained numbers depend on the critical path of the target block for each implementation.

Column 3 and 4 in Table 2 show the maximum number of glitched clock cycles to obtain enough faulty ciphertexts for each implementation. As discussed in Sects. 2.3 and 2.4, in the nibble-serial implementation of the block ciphers, each nibble is processed in one clock cycle, while in the round-serial implementation of block ciphers, all 16 nibbles are processed at the same clock cycle. Thus, in the nibble-serial implementations, the attacker is required to inject clock glitch in 16 cycles to affect all nibbles, while in round-serial implementations, injecting the glitch in one cycle can affect all nibbles.

Compared to the previous fault attacks on PRESENT, we inject more faults. The attack in [11] needs up to 150 faulty ciphertexts to retrieve the unique key. The attack proposed in [14] require 48 faulty ciphertext to retrieve the last round key of the algorithm. While the number of required fault injections in the DFIA attack is bigger compared to the mentioned previous works, we provide practical results with less restrictions of the fault model.

The previous DFA attacks on LED [12], requires a random faulty nibble to decrease the key search space to 2^8 candidates. Also, the methodology proposed in [15] is based on algebraic equations and injects a single fault to reduce the key search space to $26 \sim 217$ key guesses. The proposed DFIA attack on LED finds the unique correct key guess using additional fault injections. However, the biased fault model is practical and easy to achieve for the attacker.

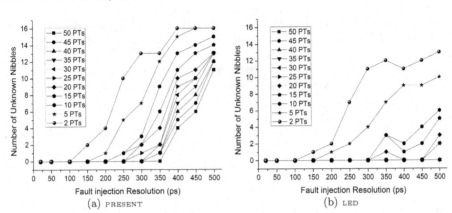

Fig. 7. Trade-off between fault injection resolution and number of plaintexts used for (a)PRESENT and (b)LED

4.2 Trade-Off Between Fault Injection Resolution and Number of Plaintexts

In this section, we provide the experimental results to verify the efficiency of the extended version of DFIA. We investigate the relationship between fault injection resolution and the number of plaintexts that DFIA needs to retrieve the key. As our fault injection means is clock glitching, our fault injection resolution is the minimum increment or decrement in the clock glitch width that we can achieve. In this experiment, we apply DFIA attacks (Algorithm 1) on our PRE-rs and LED-ns implementations for different fault injection resolutions, from 20 ps to 500 ps, and for different number of plaintexts, which ranges from 2 to 50. Then, we count the number of the key nibbles that DFIA cannot retrieve under a given fault injection resolution and a given number of plaintexts. We call such nibbles as unknown nibbles throughout this section.

Figures 7(a) and 7(b) present the results for PRE-rs and LED-rs implementations, respectively. In these figures, the Y axis shows the number of unknown nibbles out of 16 nibbles and the X axis shows clock glitch resolutions. Each data line in the graphs corresponds to a different number of applied plaintexts (PTs). Figures 7(a) and 7(b) show two important behaviors for both LED-rs and PRE-rs. For a given fault injection resolution, using more plaintexts decreases the number of unknown nibbles. For a fixed number of plaintexts, the number of unknown nibbles decreases as the fault injection resolution increases (i.e., clock glitch step size decreases). An adversary can decrease the number of unknown nibbles either by increasing the fault injection resolution or by increasing the number of plaintexts (i.e., encryptions). Therefore, we can conclude that there is a trade-off between the fault injection resolution and the number of required plaintexts. Due to this trade-off, DFIA can still efficiently retrieve the key when the fault injection equipment has a low resolution or when few biased faults are available.

5   Related Work

Ghalaty et al. [5] have a discussion on the differences of the DFIA attack with other types of attack such as DFA [4], FSA [8] and DPA [3]. In this section, we will talk about the previous types of fault attacks that use the concept of biased fault and explain their differences with DFIA. Although DFIA [5] is not the first work that utilizes the fault bias as a fault model [9,11,16,17], it is the first work that defines biased fault just beyond the fault sensitivity.

Lashermes et al. [9] assume a biased fault model and use the concept of hypothesis test on a distinguisher (i.e. Shannon entropy). However, in order for their method to converge in a practical time, they need a method to quantify the characteristic error distribution of the fault injection means. For this purpose, they need to profile the device under attack for different data sets. On the other hand, DFIA does not require any profiling phase.

Li et al. [16] and De Santis et al. [11] apply similar methodology to AES and PRESENT algorithms. They assume that faults will cause bias in the intermediate values, for example because of stuck-at faults in the intermediate value. In contrast, DFIA assumes that the fault itself is biased: The faulty intermediate values must differ in a small number of bits from the non-faulty value. But the faulty intermediate values do not have to be biased.

Jarvinen et al. [18] also propose DFA attack based on biased fault injection model. However, their method is more similar to the DFA attacks. Their definition of fault model is also different from DFIA's fault model. In this paper, the author defines the biased fault as the fact that the probability of stuck-at-1 or stuck-at-0 is higher compared to the other one. They assume that based on the fault injection method, the attacker knows the value of faulty bit and can use mathematical equation to reverse the faulty ciphertext and get the key.

6   Conclusion

In this paper, we propose a DFIA on round-serial and nibble-serial implementations of PRESENT and LED. Based on our result, we can retrieve the unique key guess for each algorithm with a reasonable number of fault injections. Our method of fault injection is the clock glitching in this paper which is a very cheap and easy way of attacking for the adversary. We also study the relation between the number of plaintexts (encryptions) used, and the resolution of the fault injection equipment.

Acknowledgment. This research was supported through the National Science Foundation Grant 1441710, Grant 1115839, and through the Semiconductor Research Corporation.

References

1. ISO: Information Technology-Security Techniques-Lightweight Cryptography-Part 2: Block Ciphers. ISO/IEC 29192–2:2012, International Organization for Standardization (2012)

2. Atzori, L., Iera, A., Morabito, G.: The internet of things: a survey. Comput. Netw. **54**, 2787–2805 (2010)
3. Kocher, P.C., Jaffe, J., Jun, B.: Differential power analysis. In: Wiener, M. (ed.) CRYPTO 1999. LNCS, vol. 1666, pp. 388–397. Springer, Heidelberg (1999)
4. Biham, E., Shamir, A.: Differential fault analysis of secret key cryptosystems. In: Kaliski Jr., B.S. (ed.) CRYPTO 1997. LNCS, vol. 1294, pp. 513–525. Springer, Heidelberg (1997)
5. Ghalaty, N.F., Yuce, B., Taha, M., Schaumont, P.: Differential Fault Intensity Analysis. In: 2014 Workshop on Fault Diagnosis and Tolerance in Cryptography (FDTC), pp. 34–43. IEEE (2014)
6. Bogdanov, A.A., Knudsen, L.R., Leander, G., Paar, C., Poschmann, A., Robshaw, M., Seurin, Y., Vikkelsoe, C.: PRESENT: an ultra-lightweight block cipher. In: Paillier, P., Verbauwhede, I. (eds.) CHES 2007. LNCS, vol. 4727, pp. 450–466. Springer, Heidelberg (2007)
7. Guo, J., Peyrin, T., Poschmann, A., Robshaw, M.: The LED Block Cipher. In: Preneel, B., Takagi, T. (eds.) CHES 2011. LNCS, vol. 6917, pp. 326–341. Springer, Heidelberg (2011)
8. Li, Y., Sakiyama, K., Gomisawa, S., Fukunaga, T., Takahashi, J., Ohta, K.: Fault sensitivity analysis. In: Mangard, S., Standaert, F.-X. (eds.) CHES 2010. LNCS, vol. 6225, pp. 320–334. Springer, Heidelberg (2010)
9. Giraud, C.: DFA on AES. In: Dobbertin, H., Rijmen, V., Sowa, A. (eds.) AES 2005. LNCS, vol. 3373, pp. 27–41. Springer, Heidelberg (2005)
10. Altera Corporation: ModelSim Altera Starter Edition. http://www.altera.com
11. De Santis, F., Guillen, O., Sakic, E., Sigl, G.: Ciphertext-only fault attacks on PRESENT. In: Third International Workshop on Lightweight Cryptography for Security and Privacy, pp. 84–105 (2014)
12. Jeong, K., Lee, C.: Differential fault analysis on block cipher LED-64. In: (Jong Hyuk) Park, J.J., Leung, V.C.M., Wang, C.-L., Shon, T. (eds.) Future Information Technology, Application, and Service. LNEE, vol. 164, pp. 747–755. Springer, Heidelberg (2012)
13. Endo, S., Sugawara, T., Homma, N., Aoki, T., Satoh, A.: An on-chip glitchy-clock generator for testing fault injection attacks. J. Cryptographic Eng. **1**, 265–270 (2011)
14. Bagheri, N., Ebrahimpour, R., Ghaedi, N.: New differential fault analysis on PRESENT. EURASIP J. Adv. Sig. Process. **2013**(1), 1–10 (2013)
15. Zhao, X.j., Guo, S., Zhang, F., Wang, T., Shi, Z., Ji, K.: Algebraic Differential Fault Attacks on LED Using a Single Fault Injection. IACR Cryptology. ePrint Archive 2012/347 (2012)
16. Li, Y., Hayashi, Y., Matsubara, A., Homma, N., Aoki, T., Ohta, K., Sakiyama, K.: Yet another fault-based leakage in non-uniform faulty ciphertexts. In: Danger, J.-L., Debbabi, M., Marion, J.-Y., Garcia-Alfaro, J., Heywood, N.Z. (eds.) FPS 2013. LNCS, vol. 8352, pp. 272–287. Springer, Heidelberg (2014)
17. Fuhr, T., Jaulmes, E., Lomné, V., Thillard, A.: Fault attacks on aes with faulty ciphertexts only. In: 2013 IEEE Workshop on Fault Diagnosis and Tolerance in Cryptography (FDTC), pp. 108–118. IEEE (2013)
18. Jarvinen, K., Blondeau, C., Page, D., Tunstall, M.: Harnessing biased faults in attacks on ECC-based signature schemes. In: 2012 IEEE Workshop on Fault Diagnosis and Tolerance in Cryptography (FDTC), pp. 72–82. IEEE (2012)

A Biased Fault Attack on the Time Redundancy Countermeasure for AES

Sikhar Patranabis[✉], Abhishek Chakraborty, Phuong Ha Nguyen, and Debdeep Mukhopadhyay

Department of Computer Science and Engineering,
IIT Kharagpur, Kharagpur, India
{sikharpatranabis,phuongha}@gmail.com,
{abhishek.chakraborty,debdeep}@cse.iitkgp.ernet.in

Abstract. In this paper we propose the first practical fault attack on the time redundancy countermeasure for AES using a biased fault model. We develop a scheme to show the effectiveness of a biased fault model in the analysis of the time redundancy countermeasure. Our attack requires only faulty ciphertexts and does not assume strong adversarial powers. We successfully demonstrate our attack on simulated data and 128-bit time redundant AES implemented on Xilinx Spartan-3A FPGA.

Keywords: Cryptanalysis · Time redundancy · Biased faults · AES

1 Introduction

Implementation attacks on secure embedded systems come in different flavors. One of these is the Side-Channel Analysis (SCA) such as *Differential Power Analysis* [8]. The other popular variety is the active Fault Analysis (FA) involving injection of faults into cryptographic systems and analysis under different fault models [2]. Attacks such as the Differential Fault Intensity Analysis (DFIA) [4] have in fact combined DPA with fault injection principles to obtain biased fault models. The advantage of a biased fault model lies in the ability of the adversary to derive an intermediate key-dependent state variable under several key hypotheses. The correct key hypothesis produces small changes to the faulty state while incorrect ones infer big, random changes.

This work attacks the time redundancy countermeasure using a biased fault model. The model is not as strict as some proposed earlier, such as stuck-at-zero or stuck-at-one faults [3]. The time redundancy technique is as an effective countermeasure, in which an encryption is followed by a redundant encryption, and in the event of a mismatch, *the faulty ciphertext is either suppressed or replaced by a random ciphertext*. Literature proposes time redundancy as a classical fault tolerance technique [10,11] with the assumption of a uniform unbiased fault distribution. For a time redundant AES, in order to obtain the faulty ciphertext, the adversary must introduce exactly the same fault in both the actual and redundant round cycles. When the fault distribution is unbiased (as classically assumed) the probability of occurrence of this event is very low. But a biased

© Springer International Publishing Switzerland 2015
S. Mangard and A.Y. Poschmann (Eds.): COSADE 2015, LNCS 9064, pp. 189–203, 2015.
DOI: 10.1007/978-3-319-21476-4_13

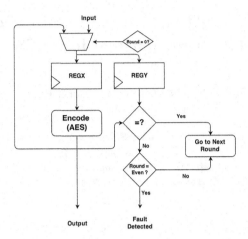

Fig. 1. Time redundancy

fault model augments this probability to the extent that it is feasible to obtain sufficient number of faulty ciphertexts to recover the key, while using a practical number of fault injections, even in the presence of the time redundancy countermeasure.

This work also assumes that we are operating only on faulty ciphertexts unlike traditional DFA which requires fault-free ciphertexts as well [1,6,9,12,14]. The proposed attack, like Differential Fault Intensity Analysis (DFIA) [4] targets an affected state variable by a biased fault injection methodology to retrieve the key. Our contributions are threefold: First, we develop a formulation for the degree of biasness in the fault distribution. Second, we propose fault models for biased faults and demonstrate actual fault attacks on a real life AES implementation with the time redundancy countermeasure. Finally, we establish through simulations and real life experiments that the number of fault injections required to defeat the time redundancy countermeasure is inversely proportional to the biasness of the fault induced.

2 Related Work

2.1 The Time Redundancy Countermeasure

Figure 1 illustrates the use of time redundancy in fault detection. Time redundancy is a fault tolerance technique that uses additional time to perform the functions of a system multiple times and compares the results to detect faults if any. A particular advantage of this approach is its low area overhead. The basic time redundancy technique has essentially three important aspects - repetition of function computation, storage of results of original and redundant computations and comparison of results for fault detection. In ciphers, time redundancy is often used for concurrent error detection(CED) against DFA by repeating each round twice and comparing the results. Previous research has proposed some countermeasures to fault attacks using time redundancy. These include re-computation [11] as well as double data rate computation [10].

2.2 Fault Attacks on AES

Recent research has focused on two broad categories of fault analysis of AES - attacks that require correct and faulty ciphertext pairs, and attacks that require faulty ciphertexts only. The first category principally includes Differential Fault Analysis(DFA). In DFA, the adversary compares the response of the cipher with and without fault injections [1,6,12,14]. The other category of fault attacks on AES require only faulty ciphertexts to retrieve the key, as proposed by Fuhr et al. [3]. The attack uses stuck-at fault models and depends on the degree of control the adversary has on the distribution of the injected fault. A very similar approach proposed by Ghalaty et al. is the Differential Fault Intensity Analysis (DFIA) [4] that uses a biased but slightly less restrictive single byte fault model. Both these approaches make several key hypotheses on the affected state bytes in order to retrieve a hypothetical value whose distribution is strongly biased.

Our proposed attack uses a biased fault model to attack the time redundancy countermeasure for AES-128 using faulty ciphertexts only. The reason is that recovering the key using only faulty ciphertexts is widely believed to be more challenging. However, similar attack procedure using biased fault models can also be developed for the former scenario since we can always obtain fault free ciphertext as well. Biased fault models expose a significant vulnerability of the classical time redundancy countermeasure. Unlike in a uniform fault distribution, a biased fault distribution implies that the adversary can introduce the same fault in both the normal and the redundant computation cycles with high probability. This reduces the number of fault injections required per faulty ciphertext. As in DFIA, our attack achieves the desired fault distribution using clock glitches at various frequencies.

3 Fault Model and Fault Injection Set up

In this section, we describe the fault model used for our attack and the fault injection set up employed to achieve this fault model.

3.1 Fault Model

Depending on the type and method of fault injection, different types of faults may occur with varying granularity such as single bit upsets, multi bit upsets, single and multi byte upsets, and diagonal upsets. Some previous works have considered *random effect on one byte*, where a single state byte may have changed to any random value [1,5,7,14]. However, such a fault model has a uniform distribution. More recent work [4] has demonstrated that single-bit, two-bit, three-bit and four-bit upsets are achievable using clock glitches, and that one can control the

Table 1. Fault model description

(a) The Fault Model

Symbol	Fault Model
FF	Fault Free
SBU	Single Bit Upset
SBDBU	Single Byte Double Bit Upset
SBTBU	Single Byte Triple Bit Upset
SBQBU	Single Byte Quadruple Bit Upset
OSB	Other Single Byte Faults
MB	Multiple Byte Faults

(b) Impact of fault location precision

Fault Model	Faults Possible(n) (Situation-1)	Faults Possible(n) (Situation-2)
SBU	8	128
SBDBU	28	448
SBTBU	56	896
SBQBU	70	1120
OSB	93	744

granularity of fault injection by varying the fault intensity. We have ourselves verified that such faults can be achieved in hardware implementations of AES-128 via introduction of clock glitches at varying frequencies (refer Sect. 3.2).

For further discussions in this paper , we distinguish between major classes of faults that covers the entire possible fault state. Table 1a summarizes these categories. Our experiments have shown that SBU is the most suitable fault model for our attacks on time-redundant AES implementations. However, we also present results for SBDBU, SBTBU and SBQBU to show the impact of fault model granularity on the performance of our attacks. Note that the degree of control that the attacker has on the fault location impacts the fault models in terms of the number of possible fault (N) under that fault model. We distinguish between the following two situations - Situation-1 when the attacker has perfect control over the faulty byte and Situation-2 when the attacker does not have control over the faulty byte.

In the case of single byte faults, if k be the number of bit upsets in the target byte, then the number of possible faults in either scenario is different. In Situation-1, any k bits of the fixed target byte is affected, so number of possible faults is $\binom{8}{k}$. In Situation-2, however k bits of any target byte could be affected, so number of possible faults is $16\binom{8}{k}$, which is 16 times greater than in Situation-1.

Table 1b captures the number of possible faults under various fault models in both situations. Evidently, precision in terms of fault location restricts the set of possible faults under a fault model significantly. Note that n is the total number of faults possible under the fault model.

3.2   Fault Injection Set up

Figure 2 describes our set up for fault injection in time redundant AES-128.

The set up consists of an FPGA (Spartan-3A XC3S400A), a PC and an external arbitrary function generator (Tektronix AFG3252). The FPGA has a DUT (Device Under Test) block, which is a time-redundant AES implementation. Faults are injected using clock glitches and the fault intensity is controlled by increasing/decreasing the glitch frequency. The system has two clock signals - clk_{slow} and clk_{fast}, derived from an external clock signal clk_{ext} via a Xilinx Digital Clock Manager (DCM) module. The clk_{ext} is generated by the external function generator and can take frequency values up to 120 MHz. The clk_{slow} signal has the same frequency as clk_{ext} and is used for fault-free operation of the DUT. The

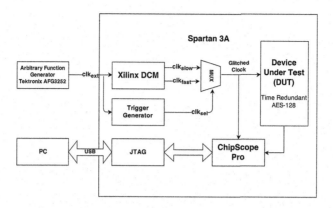

Fig. 2. Fault injection setup

clk_{fast} signal has a frequency equal to twice the frequency of clk_{ext} and is used to create the glitches for fault injection. The appropriate signal is fed to the DUT via a MUX. The select line of the MUX is the clk_{sel} signal which is output by the trigger generator and is set to high when clk_{fast} is to be fed to the DUT. The faulty states of the registers were monitored using Chipscope Pro 12.3 analyzer.

We injected faults in both the original and redundant rounds of time-redundant AES-128 by varying the clk_{ext} over a wide range of frequencies. Since the Chipscope pro 12.3 Analyzer limits the number of observable samples at a given frequency to 1024, we observed 512 samples for the original round and 512 samples for the redundant round. Tables 2a and b summarize the fault patterns obtained in either round. Table 3 summarizes the common frequency ranges between either round where each type of fault model is predominant.

4   Effectiveness of the Biased Fault Model

In this section, we demonstrate the effectiveness of the biased fault model in our attack. We quantify the biasness of a given fault model using the variance of the fault probability distribution. We assume that the set of faults that can occur under the fault model is given by $\mathcal{F} = \{f_1, \ldots, f_i, \ldots, f_n\}$, where n is the total number of faults possible under the fault model. Let F be a random variable that denotes the outcome of random occurrence of a single fault under this fault model. So the probability of occurrence of fault f_i is given by $p_i = Pr[F = f_i]$. Evidently, the fault model follows the probability distribution $\mathcal{P} = \{p_1, \ldots, p_i, \ldots, p_n\}$.

In order to get a faulty ciphertext in time redundant AES, the same fault f_i must occur in both the original and redundant rounds of computation. Let F_{org} and F_{red} be the random variables denoting the outcome of fault injections in the original and redundant rounds respectively. Since the fault injection in the original and redundant rounds are independent, we have $Pr[F_{org} = f_i, F_{red} = f_j] = p_i p_j$. We focus on the event where $F_{org} = F_{red}$. Let the probability of this event be denoted by $\tilde{p}$.

Table 2. Fault Distribution

(a) Fault Distribution Pattern - Original Round (b) Fault Distribution Pattern - Redundant round

Fast Clock Frequency (MHz)	FF	SBU	SBDBU	SBTBU	SBQBU	OSB	MB
125.0	512	0	0	0	0	0	0
125.1	503	9	0	0	0	0	0
125.2	489	22	1	0	0	0	0
125.3	456	50	6	0	0	0	0
125.4	425	59	22	6	0	0	0
125.5	396	45	43	28	0	0	0
125.6	354	34	112	32	0	0	0
125.7	303	23	101	85	0	0	0
125.8	260	11	55	86	0	0	0
125.9	208	5	46	147	6	0	0
126.0	176	1	39	228	68	0	0
126.1	143	0	18	211	136	4	0
126.2	115	0	10	94	178	15	0
126.3	101	0	8	95	251	49	8
126.4	65	0	9	45	232	141	20
126.5	32	0	5	16	131	187	141
126.6	13	0	3	8	98	101	289
126.7	5	0	1	4	32	112	358
126.8	0	0	1	2	5	105	399
126.9	0	0	1	2	5	88	421
127.0	0	0	0	1	2	33	476
127.1	0	0	0	0	1	12	499
127.2	0	0	0	0	0	0	512
127.3	0	0	0	0	0	0	512
127.4	0	0	0	0	0	0	512
127.5	0	0	0	0	0	0	512

Fast Clock Frequency (MHz)	FF	SBU	SBDBU	SBTBU	SBQBU	OSB	MB
125.0	512	0	0	0	0	0	0
125.1	512	0	0	0	0	0	0
125.2	507	5	0	0	0	0	0
125.3	479	32	1	0	0	0	0
125.4	456	50	8	4	0	0	0
125.5	416	63	29	4	0	0	0
125.6	375	41	67	29	0	0	0
125.7	345	29	120	32	0	0	0
125.8	303	23	158	28	0	0	0
125.9	255	11	121	123	2	0	0
126.0	215	3	51	251	2	0	0
126.1	192	1	39	214	66	0	0
126.2	131	0	11	187	177	25	0
126.3	105	0	10	104	278	15	0
126.4	87	0	8	64	231	98	24
126.5	50	0	8	46	157	162	90
126.6	27	0	5	16	113	125	226
126.7	21	0	4	10	98	118	261
126.8	13	0	3	6	50	103	337
126.9	7	0	3	5	21	107	369
127.0	3	0	3	2	12	91	401
127.1	2	0	1	1	8	44	456
127.2	0	0	0	1	7	17	487
127.3	0	0	0	0	3	8	501
127.4	0	0	0	0	1	3	508
127.5	0	0	0	0	0	0	512

Table 3. Fault Models and Corresponding Frequency Ranges

Fault Model	Frequency Range (Original and Redundant Rounds) (MHz)
FF	< 125.3
SBU	125.3-125.4
SBDBU	125.6-125.7
SBTBU	126.0-126.1
SBQBU	126.3-126.4
OSB	126.5
MB	> 127.2

$$\tilde{p} = \sum_{i=1}^{n} Pr[F_{org} = f_i, F_{red} = f_i] = \sum_{i=1}^{n} p_i^{2}. \qquad (1)$$

Evidently, this is also the probability of leakage of faulty ciphertexts. Our objective is to find if there is a correlation between the biased nature of the fault distribution and this probability of fault co-occurrence. Given the fault model $\mathcal{F}$ and the corresponding probability distribution $\mathcal{P}$, let Var denote that variance of $\mathcal{P}$. From the standard definition of variance of a probability distribution is given by $Var = \frac{\sum_{i=1}^{n} p_i^{2}}{n} - \frac{1}{n^2}$.

Table 4. Notations Used

P	Plaintext
C	Fault-free ciphertext
f_i	A specific fault instance
n	The number of possible faults under the fault model
N_C	The total number of faulty ciphertexts obtained (excluding random ciphertexts generated by the countermeasure)
N_F	The total number of fault injections
C'_{f_i}	The faulty ciphertext under fault f_i
r	A round of AES
k	A key hypothesis
K	The correct key
S^r_K	The fault free cipher state in round r for key K
S'^r_{k,f_i}	A guess for the faulty cipher state before the SubBytes of round r under fault f_i and key hypothesis k

Note that the value of Var is 0 for a uniform fault distribution and increases with increase in non-uniformity. This justifies using the variance of the fault probability distribution as a measure for quantifying the biasness of the fault model. Finally, we have the following relation.

$$\tilde{p} = nVar + \frac{1}{n} \tag{2}$$

Thus, by using a biased fault model, one could greatly enhance the probability of occurrence of identical faults in consecutive rounds of computation in a time redundant circuit. The significance of this is as follows:

- If the countermeasure suppresses the ciphertext on fault detection, a biased fault model will warrant much fewer fault injections to get a faulty ciphertext.
- If the countermeasure produces a random ciphertext on fault detection that does not contribute to hypothesis testing, a biased fault model will require fewer ciphertexts and hence fewer fault injections.

In either scenario, the countermeasure is weakened.

5   Description of the Attack

In this section, we describe the detailed procedure of the performed attacks on a time redundant version of AES. The attack procedure introduces the fault into either round 8 or round 9 of AES, and exploits the biased nature of the introduced fault to decipher the key.

Please refer to Table 4 for the notations used for describing the attack procedure. Note that our fault model for the attack only comprises SBU, SBDBU,

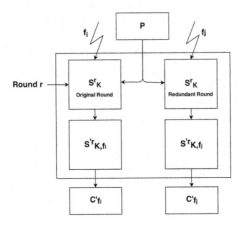

Fig. 3. Attack steps

SBTBU and SBQBU (refer Table 1a), i.e., all the fault models are *single byte fault models.*

5.1   General Attack Procedure

We now present the general steps of the attack, irrespective of the round in which the fault is introduced. A more round-specific treatment of the attack is presented following the general discussion. Table 4 summarizes the notations used in describing the attack procedure. The steps are also elucidated in Fig. 3.

Step 1: In this step the adversary induces faults f_i and f_j in both the normal and redundant computation of the target round r. However, the adversary can get the desired faulty ciphertext C'_{f_i} only if f_i and f_j are identical; otherwise the ciphertext is suppressed. *Note that alternatively, if the countermeasure produces random ciphertexts on fault detection instead of suppressing, the attack procedure does not change. The random ciphertext cannot distinguish between correct and incorrect key hypotheses and so, does not contribute to key hypothesis testing. This only increases the number of fault injections required to recover the key, as in the case of suppression.* For the purpose of a general treatment that encompasses both the scenarios, we consider N_C to be the *number of non-random faulty ciphertexts* and N_F to be the *overall number of fault injections.*

Step 2: Once the adversary collects the value of faulty ciphertext C'_{f_i}, he can compute the value of faulty state S''^r_{k,f_i} under key hypothesis k. He computes this value for every possible key hypothesis k.(Note that it is sufficient to hypothesize only those bytes of k that affect the faulty byte of S''^r_{k,f_i} since our fault model allows only single byte faults). After doing this for several collected ciphertexts, the adversary uses a distinguisher to identify the correct key hypothesis.

Step 3: The adversary chooses the key hypothesis k that minimizes/maximizes the appropriate distinguisher function for the chosen fault model. A detailed description of the distinguisher functions is presented in Sect. 5.2. If no satisfactory key guess can be made, N_C is to be increased and the test repeated. Note that in time redundant AES with suppression, the number of fault injections N_F is greater than N_C as not all fault injections yield a faulty ciphertext.

5.2 Distinguisher Functions

Distinguisher functions are used by the adversary to decide on the correct key byte(s) by selecting the key hypothesis that corresponds to the expected bias in the faulty state. For our attacks, we use two well known distinguisher functions - *Hamming Distance* [4] and *Squared Euclidean Imbalance* [3,13]. Eqs. 3 and 4 describe these functions, with k as the key hypothesis and b as the affected byte of the AES state.

$$H(k) = \sum_{i=1}^{N_C} \sum_{j=1}^{i-1} HD(S''_{k,f_i}, S''_{k,f_j}) \tag{3}$$

$$S(k) = \sum_{\delta=1}^{255} (\frac{\#\{b \mid S''_{k,f_i}[b] = \delta\}}{N_C} - \frac{1}{256})^2 \tag{4}$$

5.3 The Attack on Time Redundant AES-128

We describe the fault attack procedure where the faults are introduced in rounds 8 and 9 of AES, and the choice of distinguisher function is made accordingly.

Attack on the 8th Round

Fault Location: The fault f_i is injected just after the ante-penultimate AddRoundKey operation of the AES, modifying a random byte b of $S^8{}_K$ [3]. The injection occurs in both the original and redundant rounds of computation.

Attack Procedure: Eq. 6 summarizes the relation between the faulty ciphertext and the faulty state. The adversary can hypothesize on 4 bytes of K_{10} and one byte of K_9 to get the corresponding states and then use the SEI distinguisher to identify the correct key hypothesis, because the Hamming Distance is found to require more faulty ciphertexts in this case to arrive at the key hypothesis.

Attack Complexity: The attack requires 2^{32} key hypotheses for recovering 4 bytes of the key [3], and a total of 4 such sets for recovering the entire key, leading to an overall requirement of $4 \times 2^{32} = 2^{34}$ hypotheses. Once again, time redundancy demands that the actual number of attacks be greater than the required number of faulty ciphertexts.

$$S'^9{}_{K,f_i} = SB^{-1}(SR^{-1}(C'_{f_i} \oplus K_{10})) \tag{5}$$

$$S'^8{}_{K,f_i} = SB^{-1}(SR^{-1}((MC^{-1}((SB^{-1}(SR^{-1}(C'_{f_i} \oplus K_{10})) \oplus K_9)))) \tag{6}$$

Attack on the 9th Round

Fault Location: The fault f_i is injected just after the penultimate AddRound-Key operation of the AES, modifying a random byte b of $S^9{}_K$ [3]. The injection occurs in both the original and redundant rounds of computation.

Attack Procedure: Since the last round involves no MixColumns operation, we have Eq. 7. The adversary collects several faulty ciphertexts $C'_1, \ldots, C'_N$ on the same P and hypothesizes on one byte of the key to obtain 256 guesses of the faulty state $S'^9{}_{k,f_i}$ - one for each key hypothesis k. This is followed by the computation of $H(k)$ to identify the correct key hypothesis. It should be noted that the SEI distinguisher is useless in this context, as the distance to the uniform distribution will be the same for each hypothesis [3].

Attack Complexity: The attack requires 256 key hypotheses for recovering each byte of the key.

$$S'^9{}_{K,f_i} = SB^{-1}(SR^{-1}(C'_{f_i} \oplus K_{10})) \tag{7}$$

6 Simulated Results

In this section, we present results of simulations of attacks on AES-128 with time-redundancy countermeasure. The attack simulations were carried out on a software implementation of the time-redundant AES-128.

We divide the simulation into two major halves. In the first half, we assume the same fault for the original and redundant rounds so that each fault injection gives us a faulty ciphertext, i.e., N_C is same as N_F. Our aim here is to estimate the number of faulty ciphertexts required to recover the full key under different fault models. In the second half, we vary the probability distribution for each fault model to confirm the correlation of the bias with the number of fault injections required per faulty ciphertext, as described by Eq. 2. Here, N_C is less than N_F as the suppressions are simulated.

6.1 Simulation: Part-1

In this part of the simulation, we assume identical faults in both the original and redundant computation rounds and aim to estimate the average number of faulty ciphertexts required to recover the entire key.

In the simulation, a byte of the state at the desired attack point is chosen at random and then fault is introduced into a certain number of bits belonging to that byte, varying from 1 to 4. Note that these bits are also chosen at random. We simulate the attacks in rounds 8 and 9 respectively. In each case, the appropriate distinguisher function is used to choose the key hypothesis. Table 5 summarizes the number of faulty ciphertexts required for each fault model to guess the entire 128-bit key with 99% accuracy for the attacks on rounds 8 and 9.

Table 5. Number Of Faulty Ciphertexts Required To Guess the Entire Key With 99 % Probability

Round	Fault Model	N_C
8	SBU	320-340
	SBDBU	580-600
	SBTBU	1000-1040
	SBQBU	1900-2000
9	SBU	288-320
	SBDBU	608-640
	SBTBU	832-880
	SBQBU	1360-1440

6.2   Simulation: Part-2

In the second half of the simulation, we varied the degree of bias for each fault model by controlling the variance of the fault probability distribution for each model and observed the average number of fault injections required per faulty ciphertext, computed over a set of 100 ciphertexts. In this experiment, the assumption was that the countermeasure suppresses the ciphertext on fault injection. Our experiment considered two distinct scenarios, in which the adversary has perfect and no control respectively over the target byte in which the fault is to be induced. For the first scenario, the fault was injected only in the fixed target byte, while in second scenario, the target byte was randomly chosen. In either scenario, we simulated the fault probability variance using a normal distribution with mean $1/n$ and the desired variance, where n is the total number of faults achievable under the corresponding fault model. Figures 4a and b summarize the simulation observations over a wide range of fault distribution variances, in both scenarios. These observations show that with increase in bias of the fault distribution, the number of fault injections that are required per faulty ciphertext drops rapidly. Thus, using a fault model with high variance indeed weakens the time redundancy countermeasure.

We also simulated another experiment with the same fault model, but with the assumption that the countermeasure produces a random ciphertext instead of suppressing it. Figure 4c shows, for the attack on round 9 where the adversary has perfect control over the target byte, how the required number of fault injections varies with the variance of the probability distribution. Clearly, even in this scenario, the total number of fault injections decreases with increase in the variance of the fault probability distribution.

7   Experimental Results

In this experiment, we evaluate the proposed attack on a time-redundant hardware implementation of AES on Spartan-3A FPGA . The implementation is a

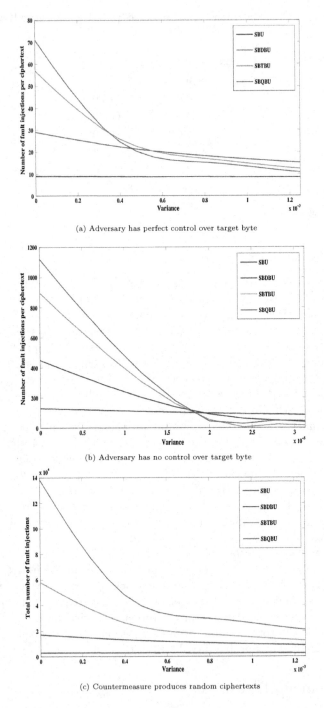

(a) Adversary has perfect control over target byte

(b) Adversary has no control over target byte

(c) Countermeasure produces random ciphertexts

Fig. 4. Number of fault attacks per faulty ciphertext vs variance of fault probability distribution

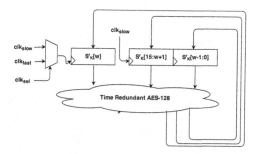

Fig. 5. Modified fault injection setup: adversary has control over affected byte

register-transfer level Verilog definition of AES with each round duplicated by a re-computation round that helps achieve time redundancy. Thus a total of 20 rounds of computation are necessary. The plaintext and key are randomly chosen 128 bit values. If the output of original and redundant round of computations is different, i.e., if a fault is detected by the countermeasure, the output is immediately suppressed.

7.1   Experimental Procedure

Attack on Round-8: A total of 4 bytes of the AES state were affected one by one after the anti-penultimate AddRoundKey operation, since each byte of the faulty state can be guessed by hypothesizing 4 bytes of the Round 10 key K_{10}. Again, the external clk_{fast} was increased gradually from 125.3 MHz to 126.4 MHz to achieve the for different fault models. Once sufficient number of faulty ciphertexts had been collected for each of the 4 bytes, the entire key was deciphered using the appropriate Squared Euclidean Imbalance computation for each byte for all the key hypotheses.

Attack on Round-9: Each of the 16 bytes of the AES state were affected one by one after the penultimate AddRoundKey operation to guess the 16 bytes of the Round 10 key K_{10}. The external clk_{fast} was increased gradually from 125.3 MHz to 126.4 MHz to achieve the four different fault models. Once sufficient number of faulty ciphertexts had been collected for each byte, the entire key was deciphered using the appropriate Hamming Weight computation for each byte for all the key hypotheses.

7.2   Fault Location Precision

We performed 2 types of attacks - Type-1 in which the adversary has perfect control over the byte in which the fault is to be introduced and Type-2 in which the adversary only knows that the fault injected is a single byte fault without any knowledge of the byte affected. The second type of experiments demands much lesser control over the actual fault injection, but is weaker as observed

Table 6. Experimental Results

Round	Fault Model	Fault Variance		N_C	N_F (simulation)		N_F (experimental)	
		Type-1	Type-2		Type-1	Type-2	Type-1	Type-2
8	SBU	9.5×10^{-2}	3.6×10^{-3}	304.75	340.48	647.52	387.67	687.91
	SBDBU	1.4×10^{-2}	9.2×10^{-4}	625.12	1456.25	1506.25	1448.45	1652.30
	SBTBU	9.7×10^{-3}	4.9×10^{-4}	1020.49	1815.60	2315.40	1974.86	2395.83
	SBQBU	3.2×10^{-3}	5.9×10^{-5}	1878.55	7868.82	28038.54	8003.14	30201.41
9	SBU	9.2×10^{-2}	3.5×10^{-3}	304.24	385.88	603.11	387.98	632.71
	SBDBU	8.8×10^{-2}	7.9×10^{-4}	624.65	641.18	1487.36	647.82	1556.69
	SBTBU	8.1×10^{-2}	6.7×10^{-4}	832.32	873.56	2054.00	878.23	2489.25
	SBQBU	7.5×10^{-2}	3.5×10^{-5}	1328.22	1788.84	17239.10	1809.25	20145.66

in the experimental results, and demands a significantly larger number of fault injections. For the first type, only the target byte should be affected by the clock glitch while in the second, the entire AES state should be subjected to the clock glitch. We describe the set up changes to be made for either scenario in greater detail. Suppose that the adversary wishes to affect only byte w of the AES state. She can achieve this precision by modifying the fault injection set up slightly to allow clk_{fast} to affect only byte w while all other bytes are driven by clk_{slow}. This ensures that in the event of a clock glitch, only byte w is affected. This is illustrated in Fig. 5. Type-2 is the normal fault injection scenario where all bytes are allowed to be affected by clk_{fast}.

For each scenario, we repeated the experiment 100 times, with the same randomly chosen key and the randomly chosen plaintext and took the average values for the number of faulty ciphertexts as well as the number of fault injections required to recover the key as well. Table 6 demonstrates the number of faulty ciphertexts and the number of fault attacks required for recovering the entire key under the attack on rounds 8 and 9, for both the scenarios where the adversary has and does not have control over the fault location. The variance of fault distribution presented for each model was experimentally observed. In both tables, we compare the experimentally required number of fault injections with the expected number of fault injections according to the simulation. It is evident that the experimentally obtained data corroborates the simulation results very well, thus confirming the hypothesis that with more bias, our proposed fault attack can break the time redundancy countermeasure with very less number of fault injections, as compared to unbiased faults.

8   Conclusions

This paper presents the first successful practical fault attack on the time redundancy countermeasure for AES-128 using biased fault models. The proposed attack requires neither precise fault injection techniques nor strong adversarial powers. The attack involves fault injection in either round 8 or round 9 of time redundant AES-128 using clock glitches. Our attack has been successfully demonstrated on simulated data as well as on 128-bit time redundant AES implemented on Xilinx Spartan-3A FPGA. The paper also develops a scheme to show the effectiveness of a biased fault model in the analysis of the time redundancy countermeasure. We conclude that the usage of the countermeasures in secure systems based on uniform fault distribution should be reconsidered in the presence of biased fault models. Our future work is to apply our proposed attack to the hardware redundancy countermeasure.

References

1. Biham, E., Shamir, A.: Differential fault analysis of secret key cryptosystems. In: Kaliski Jr, B.S. (ed.) CRYPTO 1997. LNCS, vol. 1294, pp. 513–525. Springer, Heidelberg (1997)
2. Boneh, D., DeMillo, R.A., Lipton, R.J.: On the importance of checking cryptographic protocols for faults. In: Fumy, W. (ed.) EUROCRYPT 1997. LNCS, vol. 1233, pp. 37–51. Springer, Heidelberg (1997)
3. Fuhr, T., Jaulmes, E., Lomné, V., Thillard, A.: Fault attacks on aes with faulty ciphertexts only. In: Fault Diagnosis and Tolerance in Cryptography (FDTC), pp. 108–118. IEEE (2013)
4. Ghalaty, N.F., Yuce, B., Taha, M., Schaumont, P.: Differential fault intensity analysis
5. Hemme, L.: A differential fault attack against early rounds of (Triple-)DES. In: Joye, M., Quisquater, J.-J. (eds.) CHES 2004. LNCS, vol. 3156, pp. 254–267. Springer, Heidelberg (2004)
6. Kim, C.H.: Differential fault analysis against aes-192 and aes-256 with minimal faults. In: 2010 Workshop on Fault Diagnosis and Tolerance in Cryptography (FDTC), pp. 3–9. IEEE (2010)
7. Kim, C.H.: Improved differential fault analysis on aes key schedule. IEEE Trans. Inf. Forensics Secur. $7(1)$, 41–50 (2012)
8. Kocher, P.C., Jaffe, J., Jun, B.: Differential power analysis. In: Wiener, M. (ed.) CRYPTO 1999. LNCS, vol. 1666, p. 388. Springer, Heidelberg (1999)
9. Li, Y., Sakiyama, K., Gomisawa, S., Fukunaga, T., Takahashi, J., Ohta, K.: Fault sensitivity analysis. In: Mangard, S., Standaert, F.-X. (eds.) CHES 2010. LNCS, vol. 6225, pp. 320–334. Springer, Heidelberg (2010)
10. Maistri, P., Leveugle, R.: Double-data-rate computation as a countermeasure against fault analysis. IEEE Trans. Comput. $57(11)$, 1528–1539 (2008)
11. Malkin, T., Standaert, F.-X., Yung, M.: A comparative cost/security analysis of fault attack countermeasures. In: Breveglieri, L., Koren, I., Naccache, D., Seifert, J.-P. (eds.) FDTC 2006. LNCS, vol. 4236, pp. 159–172. Springer, Heidelberg (2006)
12. Piret, G., Quisquater, J.-J.: A differential fault attack technique against SPN Structures, with application to the AES and KHAZAD. In: Walter, C.D., Koç, Ç.K., Paar, C. (eds.) CHES 2003. LNCS, vol. 2779, pp. 77–88. Springer, Heidelberg (2003)
13. Rivain, M.: Differential fault analysis on DES middle rounds. In: Clavier, C., Gaj, K. (eds.) CHES 2009. LNCS, vol. 5747, pp. 457–469. Springer, Heidelberg (2009)
14. Tunstall, M., Mukhopadhyay, D., Ali, S.: Differential fault analysis of the advanced encryption standard using a single fault. In: Ardagna, C.A., Zhou, J. (eds.) WISTP 2011. LNCS, vol. 6633, pp. 224–233. Springer, Heidelberg (2011)

Countermeasures

Faster Mask Conversion with Lookup Tables

Praveen Kumar Vadnala[✉] and Johann Großschädl

Laboratory of Algorithmics, Cryptology and Security, University of Luxembourg,
6, rue Richard Coudenhove-Kalergi, L-1359 Luxembourg, Luxembourg
{praveen.vadnala,johann.groszschaedl}@uni.lu

Abstract. Masking is an effective and widely-used countermeasure to thwart Differential Power Analysis (DPA) attacks on symmetric cryptosystems. When a symmetric cipher involves a combination of Boolean and arithmetic operations, it is necessary to convert the masks from one form to the other. There exist algorithms for mask conversion that are secure against first-order attacks, but they can not be generalized to higher orders. At CHES 2014, Coron, Großschädl and Vadnala (CGV) introduced a secure conversion scheme between Boolean and arithmetic masking of any order, but their approach requires $d = 2t + 1$ shares to protect against attacks of order t. In the present paper, we improve the algorithms for second-order conversion with the help of lookup tables so that only three shares instead of five are needed, which is the minimal number for second-order resistance. Furthermore, we also improve the first-order secure addition method proposed by Karroumi, Richard and Joye, again with lookup tables. We prove the security of all presented algorithms using well established assumptions and models. Finally, we provide experimental evidence of our improved mask conversion applied to HMAC-SHA-1. Simulation results show that our algorithms improve the execution time by 85 % at the expense of little memory overhead.

Keywords: Side-Channel Analysis (SCA) · Arithmetic masking · Boolean masking · Provably secure masking · HMAC-SHA-1

1 Introduction

Ever since the introduction of Side-Channel Analysis (SCA) attacks in the late 1990s, there has been much interest in finding countermeasures to thwart this form of "physical cryptanalysis," in particular the Differential Power Analysis (DPA) attacks [8]. From a high-level perspective, DPA countermeasures aim to either randomize the power consumption (which can be done in both the time and amplitude domain) or make it completely independent from the processed data. The goal of both approaches is to eliminate (or, at least, reduce) the correlation between the power consumption and the key-dependent intermediate variables processed during the execution of a cryptographic algorithm. Concrete examples for randomization in the time domain include various "hiding"-style countermeasures like the insertion of random delays or shuffling of operations

© Springer International Publishing Switzerland 2015
S. Mangard and A.Y. Poschmann (Eds.): COSADE 2015, LNCS 9064, pp. 207–221, 2015.
DOI: 10.1007/978-3-319-21476-4_14

[9]. On the other hand, a classical example of randomization in the amplitude domain is masking, which aims to conceal each sensitive intermediate variable x with a random value x_2, called mask [2,9]. This means that x is represented by two shares, namely the masked variable $x_1 = x \oplus x_2$ and the mask x_2. The two shares need to be manipulated separately throughout the execution of the algorithm to ensure that the instantaneous power consumption of the device does not leak any information about x. A conventional DPA attack may reveal x_1 or x_2 (both of which appear as random numbers to the attacker), but the knowledge of x_1 alone or x_2 alone does not give the attacker any information about the sensitive variable x.

One of the major challenges when applying masking to a block cipher is to implement the round functions in such a way that the shares can be processed independently from each other, while it still must be possible to recombine them at the end of the execution to get the correct result. This is fairly easy for all linear operations, but can introduce massive overheads for the non-linear parts of a cipher, i.e. the S-boxes. In addition, all round transformations need to be executed twice (namely for x_1 and for x_2, where $x = x_1 \oplus x_2$), which entails a further performance penalty. Another problem is that a basic masking scheme as described above is vulnerable to a so-called second-order DPA attack where an attacker combines information from two leakage points (i.e. he exploits the side-channel leakage originating from x_1 and x_2 simultaneously [11]). Such a second-order DPA attack can, in turn, be thwarted by second-order masking, in which each sensitive variable is concealed with two random masks and, consequently, represented by three shares. In general, a d-th order masking scheme uses d random masks to split a sensitive intermediate variable into $d + 1$ shares $x_1, x_2, \ldots, x_{d+1}$ satisfying $x_1 \oplus x_2 \oplus \cdots \oplus x_{d+1} = x$, which are then processed independently. In this way, it is guaranteed that the joint leakage of any subset of up to d shares is independent of the secret key. Only a combination of all $d + 1$ shares (i.e. the masked variable $x_1 = x \oplus x_2 \oplus \cdots \oplus x_{d+1}$ and the d masks $x_2, \ldots, x_{d+1}$) is jointly dependent on the sensitive variable. However, given the presence of noise, the cost for attacking a higher-order masked implementation increases exponentially with d [2].

Depending on the algorithmic properties of a cipher, a masking scheme can have to protect Boolean operations (e.g. xors, shifts) or arithmetic operations (e.g. modular additions). When a cipher involves both Boolean and arithmetic operations, it is necessary to convert the masks from one form to the other to obtain the correct ciphertext (or plaintext). Examples of symmetric algorithms that involve arithmetic as well as Boolean operations include the widely-used hash functions SHA-1, SHA-2, Blake and Skein, some ARX-based block ciphers (e.g. XTEA, Threefish) and all four finalists for the eSTREAM software portfolio. Given the widespread deployment of these cryptosystems in various kinds of application (including some with a need for sophisticated countermeasures against DPA), it is important to develop efficient techniques for the conversion between Boolean and arithmetic masks. However, almost all secure conversion techniques reported in the literature are only applicable to first-order masking

[4–7,10]. Among the few exceptions is the second-order conversion scheme due to Vadnala and Großschädl [13] and the recent higher-order conversion scheme by Coron, Großschädl and Vadnala [3]. We outline both schemes below.

Vadnala-Großschädl Scheme [13]. The foundation of this technique is the generic second-order countermeasure that Rivain, Dottax and Prouff proposed at FSE 2008 [12]. We recall their algorithm for computing a second-order secure masked S-box output from a second-order secure masked input below.

Algorithm 1. Sec2O-masking [12]

Input: Three input shares: $(x_1 = x \oplus x_2 \oplus x_3, x_2, x_3) \in \mathbb{F}_{2^n}$, two output shares: $(y_1, y_2) \in \mathbb{F}_{2^m}$, and an (n, m) S-box lookup function S
Output: Masked S-box output: $S(x) \oplus y_1 \oplus y_2$
1: $r \leftarrow \mathsf{Rand}(n)$
2: $r' \leftarrow (r \oplus x_2) \oplus x_3$
3: **for** $a := 0$ to $2^n - 1$ **do**
4:     $a' \leftarrow a \oplus r'$
5:     $T[a'] \leftarrow ((S(x_1 \oplus a) \oplus y_1) \oplus y_2)$
6: **end for**
7: **return** $T[r]$

In Algorithm 1, a lookup table is generated for all possible values of x. The index to the lookup table is masked using a random number r. Then, the correct value of the share is obtained by retrieving the table entry corresponding to the index r. The main idea here is that the actual computation of the third arithmetic share is hidden among other dummy calculations for all the possible values. Since the value of r changes for every iteration, the attacker is not able to guess the point in time at which the actual value of x is being leaked. The authors of [12] proved the security of the algorithm by demonstrating that no pair of intermediate variables leaks any sensitive information.

The goal of a second-order Boolean to arithmetic conversion is to compute arithmetic shares from a set of Boolean shares without introducing any second or first-order leakage. In order to achieve second-order DPA resistance, we need three Boolean shares x_1, x_2, and x_3 so that the sensitive variable x is given as $x = x_1 \oplus x_2 \oplus x_3$. The goal is to find three arithmetic shares A_1, A_2, A_3 satisfying $x = A_1 + A_2 + A_3$ without leaking any first or second-order information about x. The solution given by Vadnala and Großschädl [13] is to modify the masked lookup table in Algorithm 1 to store $((x_1 \oplus a) - A_2) - A_3$ instead of a masked S-box output; the rest of the algorithm is very similar to the original one. They followed the same approach for arithmetic to Boolean conversion.

Coron-Großschädl-Vadnala Scheme [3]. Recently, Coron, Großschädl and Vadnala proposed conversion algorithms that are secure against attacks of any order [3]. They first proposed a secure solution to add Boolean shares directly by generalizing Goubin's recursion formula [6]. Their solution has a complexity $\mathcal{O}(d^2 \cdot n)$ to secure against t-th order attacks, where $d \geq 2t + 1$ and n is the size of the masks. Then, they used this addition as subroutine to derive algorithms for conversion between Boolean and arithmetic masking, again with complexity $\mathcal{O}(d^2 \cdot n)$.

Our Contributions. The generic solution of Coron, Großschädl and Vadnala [3] requires five shares to protect against second-order attacks, which entails a significant overhead in terms of the required amount of random numbers and execution time. Although the algorithms proposed by Vadnala and Großschädl [13] require only three shares to achieve second-order resistance, they become infeasible for implementation on low-resource devices (e.g. smart cards) when $n > 10$ (the additions are performed modulo 2^n), as they require a lookup table of size 2^n.

In the present paper, we propose second-order secure conversion algorithms that overcome said limitations and can, thus, be easily applied to cryptographic constructions with arbitrary n, e.g. HMAC-SHA-1 with $n = 32$. The proposed algorithms use only three shares and are, therefore, significantly faster than the state-of-the-art. Our solution follows the basic idea of Vadnala and Großschädl (which, in turn, is based on work of Rivain, Dottax and Prouff [12]), but uses a divide and conquer approach to prevent that the lookup tables become prohibitively large. In the case of Boolean to arithmetic conversion, we divide the Boolean shares into words of $l \leq 8$ bits each and then compute the words of the corresponding arithmetic shares independently in a word-by-word fashion. Part of this procedure is to handle all the carries propagating from less to more significant words, which also need to be protected by masking to prevent any first or second-order leakage. We show that this can be achieved in an efficient and secure fashion by using separate lookup tables for the carries. Furthermore, we prove the security of our conversion schemes in the same model as [12]. Using similar techniques, we show that the efficiency of the first-order secure masked addition due to Karroumi, Richard and Joye [7] can be improved as well.

2   Efficient Second-Order Secure Boolean to Arithmetic Masking

In this section, we give the efficient Boolean to arithmetic conversion algorithm secure against attacks of second-order. The idea is to split the n-bit shares into p words (of l bits each) and convert each word independently.

2.1   Boolean to Arithmetic Masking of Second-Order

We are given three Boolean shares x_1, x_2, x_3 so that the sensitive variable x is obtained through $x = x_1 \oplus x_2 \oplus x_3$. The goal is to find three arithmetic shares

A_1, A_2, A_3 that satisfy $x = A_1 + A_2 + A_3$ without leaking any first or second-order information on x. This can be achieved by generating two shares A_2 and A_3 randomly and computing the third share as $A_1 = x - A_2 - A_3$, as done in [13] using the approach of Rivain, Dottax and Prouff from [12]. But as stated earlier, this scheme becomes infeasible for use in practice when $n > 10$ since it requires a lookup table of size 2^n. To obtain a solution for $n > 10$, we apply a divide and conquer approach. That is, we split each share into p words of l bits each and compute $(A_1^i)_{(0 \le i \le p-1)}$ independently, where $A_1 = A_1^{p-1} || \cdots || A_1^0$. In this case, we also need to handle the carries propagating from word i to word $i + 1$ properly. More precisely, these carries must be protected by masking, as otherwise they would leak information about the sensitive variable. Below, we describe our method to protect the sensitive variables along with carries and demonstrate its security with a formal proof.

We differentiate between two sets of carries: input carries (i.e. carries used for the computation of A_1^i) and output carries (i.e. carries generated while computing A_1^i). Since the computation of A_1^i involves two subtractions, there are two output carries from each word A_1^i, which become input carries for the word A_1^{i+1}. For the first word A_1^0, the input carries are initialized to 0, i.e. $c_1^0 = 0$ and $c_2^0 = 0$. We compute A_1^i from the input x^i and carries c_1^i, c_2^i as follows:

$$A_1^i = (x^i -_l c_1^i -_l A_2^i -_l c_2^i -_l A_3^i)$$

An operation of the form $a -_l b$ represents $a - b \bmod 2^l$. Similarly, the output carries c_1^{i+1}, c_2^{i+1} are computed as follows:

$$c_1^{i+1} = \mathsf{Carry}(x^i, c_1^i) \oplus \mathsf{Carry}(x^i -_l c_1^i, A_2^i) \tag{1}$$
$$c_2^{i+1} = \mathsf{Carry}(x^i -_l c_1^i -_l A_2^i, c_2^i) \oplus \mathsf{Carry}(x^i -_l c_1^i -_l A_2^i -_l c_2^i, A_3^i) \tag{2}$$

where $\mathsf{Carry}(a, b)$ represents the carry from the subtraction $a - b$. As specified by Eqs. (1) and (2), each carry computation involves two subtractions: one with the input carry (c_1^i, c_2^i) and the other with a random share (A_2^i, A_3^i). In the simplest case, a subtraction $a - b$ produces a carry when $a < b$. However, in our scenario, we have operations of the form $(a -_l c) -_l b$, whereby a and b are l-bit integers and c is either 0 or 1. In the case of $c = 0$, the above operation generates a carry if $a < b$. On the other hand, when $c = 1$, we have to take into account another case, namely $a < c$, which can only happen when $a = 0$ and $c = 1$. In this special case, the difference $a -_l c$ becomes $2^l - 1$ and a carry is generated that needs to be processed as well. However, the second subtraction can not generate a carry as $b \le 2^{l-1}$. Namely, the carries from these two cases are mutually exclusive; hence, the output carry is set to 1 when either of them produces a carry as shown in Eqs. (1) and (2). For simplicity, we define the functions $F_1 \colon \{0,1\}^{l+1} \to \{0,1\}^{l+1}$ and $F_2 \colon \{0,1\}^{2l} \to \{0,1\}^{l+1}$ as follows.

$$F_1(a, b) = a -_l b \, || \, (\mathsf{Carry}(a, b)) \tag{3}$$
$$F_2(a, b) = a -_l b \, || \, (\mathsf{Carry}(a, b)) \tag{4}$$

For a given word with index i, we can compute A_1^i as well as the output carries c_1^{i+1}, c_2^{i+1} using F_1 and F_2 according to the following equations:

$$(B_1^i||d_1^i) = F_1(x^i, c_1^i)$$
$$(B_2^i||d_2^i) = F_2(B_1^i, A_2^i)$$
$$(B_3^i||d_3^i) = F_1(B_2^i, c_2^i)$$
$$(B_4^i||d_4^i) = F_2(B_3^i, A_3^i)$$

where $A_1^i = B_4^i$ and $c_1^{i+1} = d_1^i \oplus d_2^i$, $c_2^{i+1} = d_3^i \oplus d_4^i$. As pointed out in [12], the S-box in Rivain, Dottax and Prouff's scheme must be balanced in order to be secure[1]. In our case, the function F_1 plays the same role and is balanced; consequently, the security guarantee is preserved. We first present non-randomized version of our solution below for simplicity.

Algorithm 2. Insecure 20B→A

Input: Sensitive variable: $x = x_1 \oplus x_2 \oplus x_3$
Output: Arithmetic shares: $x = A_1 + A_2 + A_3$
1: $c_1^0, c_2^0 \leftarrow 0$ ▷ Initially carry is zero
2: **for** $i := 0$ to $p - 1$ **do**
3:      $A_2^i, A_3^i \leftarrow \mathsf{Rand}(l)$ ▷ Generate output masks randomly
4:      $(B_1^i, d_1^i) \leftarrow F_1(x^i, c_1^i)$
5:      $(B_2^i, d_2^i) \leftarrow F_2(B_1^i, A_2^i)$
6:      $(B_3^i, d_3^i) \leftarrow F_1(B_2^i, c_2^i)$
7:      $(B_4^i, d_4^i) \leftarrow F_2(B_3^i, A_3^i)$
8:      $(A_1^i, c_1^{i+1}, c_2^{i+1}) \leftarrow (B_4^i, d_1^i \oplus d_2^i, d_3^i \oplus d_4^i)$
9: **end for**
10: **return** A_1, A_2, A_3

The challenge is to implement Algorithm 2 so that it does not leak any first or second-order information about the sensitive variable x as well as the carries c_1^i, c_2^i for $0 \le i \le p - 1$. We present our solution in two parts: we fist give the algorithm to securely compute the result for one word (namely A_1^i), and then we use this as a "subroutine" to compute A_1. Our solution, given in Algorithm 3, employs a similar technique as [12] (recalled in Algorithm 1) in combination with Algorithm 2. Algorithm 3 expects as input three Boolean shares, six input carry shares (three each for the two carries), two output arithmetic shares, and four output carry shares. It returns as result the third arithmetic share and the remaining two output carry shares. Similar to Algorithm 1, we create a lookup table T for all the possible values in $[0, 2^{l+2} - 1]$. Here, l bits are used to store A_1^i and two bits for the two carries correspondingly. The rest of the algorithm is very similar to the original one, except that we have to handle two extra bits for the carry[2].

---

[1] An S-box $S : \{0,1\}^n \rightarrow \{0,1\}^m$ is said to be balanced if every element in $\{0,1\}^m$ is image of exactly 2^{n-m} elements in $\{0,1\}^n$ under S.
[2] We use different tables to store the actual value and the carries so that the security proof can be easily obtained as in [12].

Algorithm 3. Sec20B→A_Word

Input: Three input shares: $(x_1^i = x^i \oplus x_2^i \oplus x_3^i, x_2^i, x_3^i) \in \mathbb{F}_{2^l}$, Six input carry shares: $g_1^i = c_1^i \oplus g_2^i \oplus g_3^i, g_2^i, g_3^i, \ g_4^i = c_2^i \oplus g_5^i \oplus g_6^i, g_5^i, g_6^i \in \mathbb{F}_2$, Output arithmetic shares: A_2^i, A_3^i, Output carry shares: $h_1^i, h_2^i, h_3^i, h_4^i$

Output: Masked Arithmetic share: $(x^i -_l A_2^i) -_l A_3^i$ and masked output carries

1: $r_1 \leftarrow \mathsf{Rand}(l); r_2 \leftarrow \mathsf{Rand}(1); r_3 \leftarrow \mathsf{Rand}(1)$
2: $r_1' \leftarrow (r_1 \oplus x_2^i) \oplus x_3^i; r_2' \leftarrow (r_2 \oplus g_2^i) \oplus g_3^i; r_3' \leftarrow (r_3 \oplus g_5^i) \oplus g_6^i;$
3: **for** $a_1 := 0$ to $2^l - 1, a_2 := 0$ to $1, a_3 := 0$ to 1 **do**
4:      $a_1' \leftarrow a_1 \oplus r_1'; a_2' \leftarrow a_2 \oplus r_2'; a_3' \leftarrow a_3 \oplus r_3'$
5:      $(B_1^i, d_1^i) \leftarrow F_1((x_1^i \oplus a_1), (g_1^i \oplus a_2))$
6:      $(B_2^i, d_2^i) \leftarrow F_2(B_1^i, A_2^i)$
7:      $(B_3^i, d_3^i) \leftarrow F_1(B_2^i, (g_4^i \oplus a_3))$
8:      $(B_4^i, d_4^i) \leftarrow F_2(B_3^i, A_3^i)$
9:      $e_1^i \leftarrow ((d_1^i \oplus h_1^i) \oplus d_2^i) \oplus h_2^i$
10:     $e_2^i \leftarrow ((d_3^i \oplus h_3^i) \oplus d_4^i) \oplus h_4^i$
11:     $(T_1[a_1'||a_2'||a_3'], T_2[a_1'||a_2'||a_3'], T_3[a_1'||a_2'||a_3']) \leftarrow (B_4^i, e_1^i, e_2^i)$
12: **end for**
13: **return** $T_1[r_1||r_2||r_3], T_2[r_1||r_2||r_3], T_3[r_1||r_2||r_3]$

Finally, we give our second-order secure technique to obtain three arithmetic shares corresponding to the three Boolean shares in Algorithm 4. For the first word (i.e. $i = 0$), there are no input carries and, consequently, the three shares for both carries are set to zero (Step 1), i.e. we have $g_1^0 = g_2^0 = g_3^0 = c_1^0 = 0$ and $g_4^0 = g_5^0 = g_6^0 = c_2^0 = 0$. To protect the output carries, we use four uniformly generated random bits: $h_1^i, h_2^i, h_3^i, h_4^i$; two each for the two carries. The third share for the carries as well as A_1^i are computed recursively using the function Sec20B→A_Word (Algorithm 3)[3]. Note that for word i, $g_1^i \oplus g_2^i \oplus g_3^i = c_1^i$ and $g_4^i \oplus g_5^i \oplus g_6^i = c_2^i$. The time complexity of the overall solution is $\mathcal{O}(2^{l+2} \cdot p)$ and the memory requirements amount to $(2^{l+2} \cdot (l+2))$ bits.

Algorithm 4. Sec20B→A

Input: Boolean shares: $x_1 = x \oplus x_2 \oplus x_3, x_2, x_3$
Output: Arithmetic shares: A_1, A_2, A_3 so that $x = A_1 + A_2 + A_3$
1: $g_1^0, g_2^0, g_3^0, g_4^0, g_5^0, g_6^0 \leftarrow 0$ ▷ Initially carry is zero
2: **for** $i := 0$ to $p - 1$ **do**
3:      $A_2^i, A_3^i \leftarrow \mathsf{Rand}(l)$ ▷ Generate output masks randomly
4:      $h_1^i, h_2^i, h_3^i, h_4^i \leftarrow \mathsf{Rand}(1)$
5:      $(A_1^i, g_1^{i+1}, g_4^{i+1}) \leftarrow$ Sec20B→A_Word $((x_j^i)_{1 \leq j \leq 3}, (g_j^i)_{1 \leq j \leq 6}, A_2^i, A_3^i, (h_j^i)_{1 \leq j \leq 4})$
6:      $g_2^{i+1}, g_3^{i+1}, g_5^{i+1}, g_6^{i+1} \leftarrow h_1^i, h_2^i, h_3^i, h_4^i$
7: **end for**
8: **return** A_1, A_2, A_3

---

[3] Every call to the function Sec20B→A_Word creates a new table and is useful for that particular word only. Hence, unlike the original method in [12], we do not reuse the table.

2.2 Security Analysis

For an algorithm to be secure against second-order DPA attacks, no pair of intermediate variables appearing in the algorithm should jointly leak the sensitive variable. In [12], the authors prove the security by enumerating all the possible pairs of intermediate variables and showing that the joint distribution of none of these pairs is dependent on the distribution of the sensitive variable. We use a similar method to prove the security of Algorithm 3. Thereafter, we prove the security of Algorithm 4 through induction.

Lemma 1. *Algorithm 3 is secure against second-order DPA.*

Proof. We list all intermediate variables used in Algorithm 1 and Algorithm 3 in Table 1. The intermediate variables computed through a similar technique appear in the same row. The only difference is that we have three intermediate variables instead of one for each row[4]. Hence, the security of Algorithm 3 can be derived from the same arguments as in the case of Algorithm 1.   □

Table 1. Comparison of intermediate variables used in Algorithm 1 and Algorithm 3

Intermediate variables in Algorithm 1	Intermediate variables in Algorithm 3
x_2	x_2^i, g_2^i, g_5^i
x_3	x_3^i, g_3^i, g_6^i
y_1	A_2^i, h_1^i, h_3^i
y_2	A_3^i, h_2^i, h_4^i
r	r_1, r_2, r_3
$x_2 \oplus r$	$x_2^i \oplus r_1, g_2^i \oplus r_2, g_5^i \oplus r_3$
$x_2 \oplus r \oplus x_3$	$x_2^i \oplus r_1 \oplus x_3^i, g_2^i \oplus r_2 \oplus g_3^i, g_5^i \oplus r_3 \oplus g_5^i$
a	a_1, a_2, a_3
$a \oplus r \oplus x_2 \oplus x_3$	$a_1 \oplus r_1', a_2 \oplus r_2', a_3 \oplus r_3'$
$x_1 = x \oplus x_2 \oplus x_3$	$x_1^i = x^i \oplus x_2^i \oplus x_3^i, g_1^i = c_1^i \oplus g_2^i \oplus g_3^i, g_4^i = c_2^i \oplus g_3^i \oplus g_6^i$
$x_1 \oplus a$	$x_1^i \oplus a, g_1^i \oplus a_2, g_4^i \oplus a_3$
$S(x_1 \oplus a)$	$(B_1^i \| d_1^i) = F_1((x_1^i \oplus a), g_1^i \oplus a_2)$ $(B_3^i \| d_3^i) = F_1((x_1^i \oplus a) -_l g_1^i \oplus a_2 -_l A_2^i, g_4^i \oplus a_3)$ $d_2^i = \mathsf{Carry}((x_1^i \oplus a) -_l (g_1^i \oplus a_2), A_2^i)$ $d_4^i = \mathsf{Carry}((x_1^i \oplus a) -_l (g_1^i \oplus a_2) -_l A_2^i -_l (g_4^i \oplus a_3), A_3^i)$
$S(x_1 \oplus a) \oplus y_1$	$B_2^i = (x_1^i \oplus a) -_l (g_1^i \oplus a_2) -_l A_2^i,$ $d_1^i \oplus h_1^i \oplus d_2^i, d_3^i \oplus h_3^i \oplus d_4^i$
$S(x_1 \oplus a) \oplus y_1 \oplus y_2$	$B_4^i = (x_1^i \oplus a) -_l (g_1^i \oplus a_2) -_l A_2^i -_l (g_4^i \oplus a_3) -_l A_3^i,$ $d_1^i \oplus h_1^i \oplus d_2^i \oplus h_2^i, d_3^i \oplus h_3^i \oplus d_4^i \oplus h_4^i$
$S(x) \oplus y_1 \oplus y_2$	$x^i -_l c_1^i -_l A_2^i -_l c_2^i -_l A_3^i,$ $c_1^{i+1} \oplus h_1^i \oplus h_2^i, c_2^{i+1} \oplus h_3^i \oplus h_4^i$

Theorem 1. *Algorithm 4 is secure against second-order DPA.*

Proof. To prove the security of Algorithm 4, we apply mathematical induction on the number of words p. When $p = 1$, we already know that the algorithm is secure due to the proof of Lemma 1. Now, assume that the algorithm is secure

---

[4] The only exception is for the row $S(x_1 \oplus a)$, where we have four variables.

for $p = n$. Let E_i be the set that represents the collection of all intermediate variables corresponding to the word i. Then, by the induction hypothesis, the set $\{E_1, \cdots E_n\} \times \{E_1, \cdots E_n\}$ is independent of the sensitive variables x, c_1 and c_2. For the algorithm to be secure when $p = n + 1$, the set $\{E_1, \cdots E_n, E_{n+1}\} \times \{E_1, \cdots E_n, E_{n+1}\}$ should be independent of the sensitive variables x, c_1 and c_2. Without loss of generality, we divide this set into three subsets as follows: $\{E_{n+1} \times E_{n+1}\}$, $\{E_1, \cdots E_n\} \times \{E_1, \cdots E_n\}$, and $\{E_{n+1}\} \times \{E_1, \cdots E_n\}$. The security of $\{E_{n+1} \times E_{n+1}\}$ can be established directly from the base case, and the security of $\{E_1, \cdots E_n\} \times \{E_1, \cdots E_n\}$ follows from the induction hypothesis (see above). All the intermediate variables in E_{n+1} fall into two categories: (i) the variables that are generated randomly and are independent of any variables in $\{E_1, \cdots E_n\}$, and (ii) the variables that are a function of one or more of the following: x^{n+1}, c_1^{n+1}, c_2^{n+1}. Any pair of intermediate variables involving the former category is independent of the sensitive variables by definition and the first-order resistance of the set $\{E_1, \cdots E_n\}$. The two carry shares for the word $n + 1$, namely $(c_i^{n+1})_{1 \leq i \leq 2}$, are computed from the word n. Thus, the security of $(c_i^{n+1})_{1 \leq i \leq 3} \times \{E_1, \cdots E_n\}$ is already established in $\{E_n\} \times \{E_1, \cdots E_n\}$. One can easy see that the set $(x^{n+1}) \times \{E_1, \cdots E_n\}$ is independent of any sensitive variable. Hence, the set $\{E_{n+1}\} \times \{E_1, \cdots E_n\}$ is also independent of any sensitive variable, which proves the theorem. $\qquad\square$

3   Efficient Second-Order Secure Arithmetic to Boolean Masking

In arithmetic to Boolean conversion, the problem is to find three shares x_1, x_2, x_3 satisfying $x = x_1 \oplus x_2 \oplus x_3$, where the sensitive variable x is represented by three arithmetic shares A_1, A_2, A_3 with $x = A_1 + A_2 + A_3$. To solve this problem, we follow the same strategy as in Section 2.1. We generate two Boolean shares x_2 and x_3 randomly, and compute the third share by using the relation $x_1 = ((A_1 + A_2 + A_3) \oplus x_2 \oplus x_3)$, without leaking the value of x to first or second-order DPA. We use the following approach: we first obtain a method to convert a single arithmetic share word; then we apply this procedure recursively to all the words. For each word, we have to deal with two carries corresponding to the two additions, i.e., the carry from the addition of the shares corresponding to A_2, A_3 and its subsequent addition with A_1. Our solution is described in Algorithm 5 and Algorithm 6 .

Algorithm 5 gives the solution for converting one word of Boolean shares to corresponding arithmetic shares. We again use the technique from Algorithm 1 as in Algorithm 3. As the input shares here are masked using arithmetic masking instead of Boolean masking, we have to modify the operations accordingly. Hence, the computation of r_1' (Step 2) and a_1' (Step 5) are replaced with additive operations. However, we can still mask the carries using Boolean masking as previously and hence the corresponding operations do not change (Step 3, Step 7). We create a table for all possible values in $[0, 2^{l+2} - 1]$, where l bits are used for x_1^i and the extra two bits for the carries. From $a_1' = a_1 -_l r_1'$, we have

---

Algorithm 5. Sec20A→B_Word

Input: Three input shares: $(A_1^i = (x^i - A_2^i) - A_3^i, A_2^i, A_3^i) \in \mathbb{F}_{2^l}$, Six input carry shares: $g_1^i = c_1^i \oplus g_2^i \oplus g_3^i, g_2^i, g_3^i, g_4^i = c_2^i \oplus g_5^i \oplus g_6^i, g_5^i, g_6^i \in \mathbb{F}_2$, Output Boolean shares: x_2^i, x_3^i, Output carry shares: $h_1^i, h_2^i, h_3^i, h_4^i$

Output: Third Boolean share: $x_1^i = x^i \oplus x_2^i \oplus x_3^i$ and masked output carries

1: $r_1 \leftarrow \mathsf{Rand}(l); r_2 \leftarrow \mathsf{Rand}(1); r_3 \leftarrow \mathsf{Rand}(1)$
2: $r_1' \leftarrow (A_2^i - r_1) + A_3^i$ ▷ Mask two arithmetic shares
3: $r_2' \leftarrow (r_2 \oplus g_2^i) \oplus g_3^i; r_3' \leftarrow (r_3 \oplus g_5^i) \oplus g_6^i$
4: **for** $a_1 := 0$ to $2^l - 1$ **do**
5:     $a_1' \leftarrow a_1 -_l r_1'$ ▷ $a_1' = r_1 \Longrightarrow a_1 = r_1 +_l ((A_2^i - r_1) + A_3^i)$
6:     **for** $a_2 := 0$ to $1, a_3 := 0$ to 1 **do**
7:         $a_2' \leftarrow a_2 \oplus r_2'; a_3' \leftarrow a_3 \oplus r_3'$
8:         $(B_1^i || d_2^i) \leftarrow F_3(A_1^i + (a_3 \oplus g_4^i) + ((a_2 \oplus g_1^i) +_l a_1))$
9:         $d_1^i \leftarrow \mathsf{Carry}(a_1, r_1') \oplus \mathsf{Carry}(a_1, (-(a_2 \oplus g_1^i)))$
10:        $x_1^i \leftarrow (B_1^i \oplus x_2^i) \oplus x_3^i$ ▷ Apply Boolean masking to the result
11:        $e_1^i \leftarrow (d_1^i \oplus h_1^i) \oplus h_2^i$ ▷ Apply masking to the carries
12:        $e_2^i \leftarrow (d_2^i \oplus h_3^i) \oplus h_4^i$
13:        $T_1[a_1' || a_2' || a_3'], T_2[a_1' || a_2' || a_3'], T_3[a_1' || a_2' || a_3'] \leftarrow (x_1^i, e_1^i, e_2^i)$
14:    **end for**
15: **end for**
16: **return** $T_1[r_1 || r_2 || r_3], T_2[r_1 || r_2 || r_3], T_3[r_1 || r_2 || r_3]$

$a_1 = a_1' +_l r_1'$. However, $a_1 - r_1'$ could generate a carry, which needs to be taken care while computing x_1^i. Hence, we add the previous carry $(a_2 \oplus g_2^i)$ to a_1 to get $A_2^i +_l A_3^i$ as follows:

$$a_1 +_l (a_2 \oplus g_1^i) = (r_1 +_l ((A_2^i - r_1) + A_3^i) +_l c_1^i) = A_2^i +_l A_3^i$$

when $a_1' = r_1$ and $a_2' = r_2$. The out carry d_1^i (which becomes c_1^{i+1} for the next word) can occur in two scenarios: when $a_1 < r_1'$ or when $(a_1 + (a_2 \oplus g_1^i)) \geq 2^l$ (Step 9). It is easy to see that these two cases are mutually exclusive. Now to compute x_1^i, we use function $F_3 : \{0,1\}^{l+1} \to \{0,1\}^{l+1}$, which is defined as:

$$F_3(a) = a \mod 2^l || \mathsf{Carry}(2^l, a)$$

We then call F_3 with $(A_1^i + (a_3 \oplus g_4^i) + ((a_2 \oplus g_1^i) +_l a_1))$ where a_3 represents two shares of the second carry. In this case, the first part returned by F_3 gives x^i, and the second part corresponds to the second carry which becomes c_2^{i+1} for the next word [5]. Namely, when $a_1' = r_1$, $a_2' = r_2$ and $a_3' = r_3$ we have:

$$F_3(A_1^i + (a_3 \oplus g_4^i) + ((a_2 \oplus g_1^i) +_l a_1)) = (x^i + (a_3 \oplus g_4^i)) \mod 2^l ||$$
$$\mathsf{Carry}(2^l, (x^i + (a_3 \oplus g_4^i)))$$

---

[5] Note here that even though x^i and the carries are computed in clear, they are hidden among $2^{l+2} - 1$ dummy computations, which is the main basis for Rivain et al's original algorithm.

Once we have x^i and the carries d_1^i, d_2^i, we can simply apply Boolean masks on them to obtain x_1^i and the masked carries (Steps 10, 11 and 12).

Finally we give the full algorithm to convert from arithmetic to Boolean masking in Algorithm 6. It is similar to Algorithm 4 except that the Boolean shares and arithmetic shares are interchanged.

Algorithm 6. Sec20A→B

Input: Arithmetic shares: $A_1 = x - A_2 - A_3, A_2, A_3$
Output: Boolean shares: x_1, x_2, x_3 so that $x = x_1 \oplus x_2 \oplus x_3$
1: $g_1^0, g_2^0, g_3^0, g_4^0, g_5^0, g_6^0 \leftarrow 0$ ▷ Initially carry is zero
2: **for** $i := 0$ to $p - 1$ **do**
3:      $x_2^i, x_3^i \leftarrow \mathsf{Rand}(l)$ ▷ Generate output masks randomly
4:      $h_1^i, h_2^i, h_3^i, h_4^i \leftarrow \mathsf{Rand}(1)$
5:      $(x_1^i, g_1^{i+1}, g_4^{i+1}) \leftarrow \mathsf{Sec20A{\to}B\_Word}\left((A_j^i)_{1 \leq j \leq 3}, (g_j^i)_{1 \leq j \leq 6}, x_2^i, x_3^i, (h_j^i)_{1 \leq j \leq 4}\right)$
6:      $g_2^{i+1}, g_3^{i+1}, g_5^{i+1}, g_6^{i+1} \leftarrow h_1^i, h_2^i, h_3^i, h_4^i$
7: **end for**
8: **return** x_1, x_2, x_3

Theorem 2. *Algorithm 6 is secure against second-order DPA.*

Proof. The proof of Algorithm 6 can be obtained similar to Algorithm 4 and is omitted.

4   Efficient First-Order Secure Masked Addition

This paper considers the general problem of dealing with arithmetic operations on Boolean masks. Till now, we solved this problem by converting the Boolean masks to arithmetic masks. The basic idea is that, once we have the arithmetic masks, we can perform any arithmetic operation directly and then convert the result back to Boolean masks. But there also exists an alternative approach to the original problem, namely to perform an arithmetic operation (e.g. addition) directly on Boolean masks. This idea was first studied for first-order masking in [1] and then detailed in [7]. In this section, we provide a more efficient method using lookup tables based on the conversion technique by Debraize [5].

The problem here can be described as follows: we are given Boolean shares of two n-bit sensitive variables x: x_1, r and y: y_1, s. We need to compute z_1 so that $z_1 \oplus r \oplus s = x + y$, without any first-order leakage of x and y. To achieve this, we follow the same divide-and-conquer strategy we used in Sects. 2 and 3. Namely, we divide n-bit shares into p words of l-bit each and perform addition on the words independently. Furthermore, our method also masks the carry from word i to word $i + 1$. The addition of each word is carried out with the help of a lookup table, which can be reused for all the words[6].

---

[6] We use different tables in the case of second-order masking, but we can re-use the table for first-order masking.

Our method to generate the lookup table is given in Algorithm 7. It creates a table with 2^{2l+1} entries, each requiring $l + 1$ bit of memory. Here, $2l$ bits are used for two l-bit inputs x^i, y^i and one bit for the input carry. The output consists of l-bit z^i and one bit carry. We run through all the possible 2^{2l+1} values and store the masked value of sum and carry in the lookup table. Note that the inputs masks are t_1, t_2 and ρ (carry), and out masks are t_1 and ρ (carry).

Algorithm 7. GenTable

Input:
Output: Table T, t_1, t_2, ρ
1: $t_1, t_2 \leftarrow \mathsf{Rand}(l)$; $\rho \leftarrow \mathsf{Rand}(1)$
2: **for** $A = 0$ to $2^l - 1$ **do**
3:      **for** $B = 0$ to $2^l - 1$ **do**
4:          $T[\rho||A||B] \leftarrow ((A \oplus t_1) + (B \oplus t_2)) \oplus (\rho||t_1)$
5:          $T[\rho \oplus 1||A||B] \leftarrow ((A \oplus t_1) + (B \oplus t_2) + 1) \oplus (\rho||t_1)$
6:      **end for**
7: **end for**
8: **return** T, t_1, t_2, ρ

The full technique to compute addition on Boolean shares is given in Algorithm 8. Initially, the carry is zero, which is masked with the carry mask ρ from Algorithm 7. We distinguish between carry and no-carry cases as follows: when $\beta = \rho$, then there is no carry; otherwise, $\beta = \rho \oplus 1$. Before accessing the lookup table, we change the input masks to t_1 and t_2 (step 3, 4). After we obtain the masked sum, we change the mask back to $r^i \oplus s^i$ from t_1 (step 6). Finally, the output can be obtained as $z_1 = z_1^{p-1}||\cdots||z_1^0 = (x + y) \oplus r \oplus s$.

Algorithm 8. Sec10A

Input: $x_1 = x \oplus r, r, y_1 = y \oplus s, s, T, t_1, t_2, \rho$
Output: $z_1 = (x + y) \oplus r \oplus s$
1: $\beta \leftarrow \rho$
2: **for** $i = 0$ to $p - 1$ **do**
3:      $x_1^i \leftarrow x_1^i \oplus t_1 \oplus r^i$
4:      $y_1^i \leftarrow y_1^i \oplus t_2 \oplus s^i$
5:      $(\beta||z_1^i) \leftarrow T[\beta||x_1^i||y_1^i]$
6:      $z_1^i \leftarrow (z_1^i \oplus r^i \oplus s^i) \oplus (t_1)$
7: **end for**
8: **return** z_1

Lemma 2. *Algorithm 8 is secure against first-order DPA.*

Proof. It is easy to see that the distribution of all the intermediate variables in Algorithm 8 is independent of the sensitive variables x and y. Consequently, the proof is straightforward. $\qquad\square$

5   Implementation Results

We implemented all the proposed algorithms in ANSI C and executed them on a 32-bit ARM microcontroller. The results are summarized in Table 2. We used three different word sizes (namely $l = 1, 2, 4$) for the second-order conversion algorithm and a word size of $l = 4$ for first-order masked addition[7]. In order to compare our results with that of existing techniques, we also implemented the Coron-Großschädl-Vadnala (CGV) method [3] for second-order conversion and the Karroumi-Richard-Joye (KRJ) method [7] for first-order secure addition. As expected, the improvement in case of the second-order conversion algorithms is significant due to the reduction of the number of shares from five to three. We notice that the conversion algorithms perform best when $l = 2$. Our Boolean to arithmetic conversion algorithm with negligible memory requirements (between 8 and 64 bytes) is some 86 % faster than the CGV method. Similarly, our arithmetic to Boolean conversion algorithm improves the running time by 83 %, with equivalent memory footprint. On the other hand, we improve the performance of the first-order algorithms by roughly 20 %.

Table 2. Implementation results for $n = 32$ on a 32-bit microcontroller. The column Time specifies the running time in clock cycles, rand gives the number of calls to the random number generator function, while column l and Memory refer to the word size and memory (in bytes) required for the table-based algorithms.

Algorithm	l	Time	Memory	rand
second-order conversion				
Algorithm 4	1	12186	8	226
Algorithm 4	2	11030	16	114
Algorithm 4	4	19244	64	58
Algorithm 6	1	10557	8	226
Algorithm 6	2	9059	16	114
Algorithm 6	4	15370	64	58
CGV $A \to B$ [3]	-	54060	-	484
CGV $B \to A$ [3]	-	81005	-	822
first-order addition				
KRJ addition [7]	-	371	-	1
Algorithm 8	4	294	512	3

To study the implications of our new techniques in practice, we applied them to secure HMAC-SHA-1. The achieved results are summarized in Table 3. We can see that, in the best case scenario (i.e. $l = 2$), our new algorithms perform 85 % better than the existing approaches. In the case of first-order masking, the

---

[7] We observed that, for $l < 4$, the algorithm from [7] performs better than ours.

Table 3. Running time (in thousands of clock cycles) and penalty factor compared to the unmasked HMAC-SHA-1 implementation

Algorithm	l	Time	PF
HMAC-SHA-1	-	104	1
second-order conversion			
Algorithm 4, 6	1	9715	95
Algorithm 4, 6	2	8917	85
Algorithm 4, 6	4	15329	147
CGV [3]	-	62051	596
first-order addition			
KRJ addition [7]	-	328	3.1
Algorithm 8	4	308	2.9

improvement amounts to roughly 6 %, taking into account the pre-computation time spent on the generation of the table.

6   Conclusions

In this paper, we presented new time-memory trade-off solutions for conversion between Boolean and arithmetic masking for first and second-order. In the case of second-order conversion, we reduced the number of required shares from five to three compared to the CGV method. We demonstrated that, with negligible memory consumption (up to 64 bytes), we can improve the performance of the existing algorithms by up to 85 %.

An open research problem is to find a way to perform additions on Boolean shares directly that is secure against attacks of second-order. We can not apply the generic method of [12] in this case since the S-box is not balanced. Such an S-box would require an input of size $2l + 1$ bits (i.e. l bits for each of the two arguments to add and one bit for input carry) and output the $(l + 1)$-bit sum including the carry. For this function to be balanced, each of the 2^{l+1} possible outputs must be an image of exactly 2^l elements. However, this is not the case and, consequently, a second-order attack can be mounted. Finding a solution to this problem could further improve the efficiency of second-order masking.

References

1. Beak, Y.-J., Noh, M.-J.: Differetial power attack and masking method. Trends Math. **8**(1), 53–67 (2005)
2. Chari, S., Jutla, C.S., Rao, J.R., Rohatgi, P.: Towards Sound Approaches to Counteract Power-Analysis Attacks. In: Wiener, M. (ed.) CRYPTO 1999. LNCS, vol. 1666, pp. 398–412. Springer, Heidelberg (1999)

3. Coron, J.-S., Großschädl, J., Vadnala, P.K.: Secure conversion between boolean and arithmetic masking of any order. In: Batina, L., Robshaw, M. (eds.) CHES 2014. LNCS, vol. 8731, pp. 188–205. Springer, Heidelberg (2014)
4. Coron, J.-S., Tchulkine, A.: A new algorithm for switching from arithmetic to boolean masking. In: Walter, C.D., Koç, Ç.K., Paar, C. (eds.) CHES 2003. LNCS, vol. 2779, pp. 89–97. Springer, Heidelberg (2003)
5. Debraize, B.: Efficient and provably secure methods for switching from arithmetic to boolean masking. In: Prouff, E., Schaumont, P. (eds.) CHES 2012. LNCS, vol. 7428, pp. 107–121. Springer, Heidelberg (2012)
6. Goubin, L.: A sound method for switching between boolean and arithmetic masking. In: Koç, Ç.K., Naccache, D., Paar, C. (eds.) CHES 2001. LNCS, vol. 2162, pp. 3–15. Springer, Heidelberg (2001)
7. Karroumi, M., Richard, B., Joye, M.: Addition with blinded operands. In: Prouff, E. (ed.) COSADE 2014. LNCS, vol. 8622, pp. 41–55. Springer, Heidelberg (2014)
8. Kocher, P.C., Jaffe, J., Jun, B.: Differential power analysis. In: Wiener, M. (ed.) CRYPTO 1999. LNCS, vol. 1666, pp. 388–397. Springer, Heidelberg (1999)
9. Mangard, S., Oswald, E., Popp, T.: Power Analysis Attacks - Revealing the Secrets of Smart Cards. Springer, New York (2007)
10. Neiße, O., Pulkus, J.: Switching blindings with a view towards IDEA. In: Joye, M., Quisquater, J.-J. (eds.) CHES 2004. LNCS, vol. 3156, pp. 230–239. Springer, Heidelberg (2004)
11. Oswald, E., Mangard, S., Herbst, C., Tillich, S.: Practical second-order DPA attacks for masked smart card implementations of block ciphers. In: Pointcheval, D. (ed.) CT-RSA 2006. LNCS, vol. 3860, pp. 192–207. Springer, Heidelberg (2006)
12. Rivain, M., Dottax, E., Prouff, E.: Block ciphers implementations provably secure against second order side channel analysis. In: Nyberg, K. (ed.) FSE 2008. LNCS, vol. 5086, pp. 127–143. Springer, Heidelberg (2008)
13. Vadnala, P.K., Großschädl, J.: Algorithms for switching between boolean and arithmetic masking of second order. In: Gierlichs, B., Guilley, S., Mukhopadhyay, D. (eds.) SPACE 2013. LNCS, vol. 8204, pp. 95–110. Springer, Heidelberg (2013)

Towards Evaluating DPA Countermeasures for KECCAK on a Real ASIC

Michael Muehlberghuber[1], Thomas Korak[2]([✉]), Philipp Dunst[2], and Michael Hutter[3]

[1] Integrated Systems Laboratory (IIS), ETH Zurich,
Gloriastrasse 35, 8092 Zurich, Switzerland
mbgh@iis.ee.ethz.ch

[2] Institute for Applied Information Processing and Communications (IAIK),
Graz University of Technology,
Inffeldgasse 16a, 8010 Graz, Austria
thomas.korak@iaik.tugraz.at, p.dunst@gmx.net

[3] Cryptography Research,
425 Market Street, San Francisco, CA 94105, USA
michael.hutter@cryptography.com

Abstract. We present ZORRO, a taped-out ASIC hosting three distinct authenticated encryption architectures based on the SPONGEWRAP construction. All designs target resource-constrained environments such as smart cards or embedded devices and therefore, have been protected against DPA attacks while keeping low-area as the most important design goal in mind. Each of the three architectures contains masking and hiding countermeasures. They solely differ with regard to the implemented secret-sharing scheme. While the first design is based on a 3-share threshold implementation (TI), which does not fulfill the uniformity property, the other two make use of the 3-share approach with re-masking and the 4-share approach as proposed by Bilgin et al. Our smallest, provable first-order DPA secure KECCAK implementation requires only 14.5 kGE (which is less than half of the size of related work) and contains both front-end and back-end design overheads. Moreover, we present first DPA results of the ZORRO ASIC by comparing hiding and masking countermeasures. We were able to recover the cipherkey from a masking-secured TI implementation based on three shares with about 70 000 power traces.

Keywords: Duplex construction · SPONGEWRAP · Threshold implementation · Side-channel attacks · DPA · Low-area hardware · ASIC

1 Introduction

Confidentiality and authenticity of data are among the most important cryptographic services required to transfer data securely over public communication channels. The former is commonly achieved by symmetric encryption algorithms

---

This work was done while Michael Hutter was with Graz University of Technology.

S. Mangard and A.Y. Poschmann (Eds.): COSADE 2015, LNCS 9064, pp. 222–236, 2015.
DOI: 10.1007/978-3-319-21476-4_15

while the latter is often obtained by message authentication codes (MACs). These cryptographic primitives have been treated independently in the past, which led to inefficient solutions and severe security problems [9,11]. For this reason, researchers have started to develop new hybrid algorithms that offer the desired service of authenticated encryption (AE), for instance, as part of the on-going CAESAR competition [1].

The SPONGEWRAP construction [3] uses the underlying permutation of KEC-CAK—the winner of the NIST SHA-3 competition [6] in 2012—in order to realize an AE system. Implementations using KECCAK-f in a keyed mode[1] such as SPONGEWRAP, necessarily require protection against implementation attacks such as Differential Power Analysis (DPA) [15]. Since especially smart cards and embedded systems are usually accessible by a broader mass of people, countermeasures like *hiding* or *masking* techniques are mandatory for such devices nowadays. The authors of KECCAK proposed to implement a secret sharing technique to protect keyed KECCAK instances [2,7]. This technique is based on the idea to divide key-dependent intermediate values into unique parts (so-called shares) and to re-combine them after the processing. In order to achieve first-order DPA resistance, this sharing needs to fulfill three properties: correctness, non-completeness, and uniformity [17]. Interestingly, Bilgin et al. [8] reported that the implementation in [2,7] does not fulfill the uniformity property and is therefore not provable secure against first-order DPA attacks. As a countermeasure, they proposed to inject fresh random bits in a 3-share implementation or to add an additional share (4-share version) that avoids the need of fresh randomness.

This work presents first results of an actually "taped-out" application-specific integrated circuit (ASIC), called ZORRO (our chip is not to confuse with the block-cipher of Gérard et al. [12] proposed at CHES 2013). ZORRO hosts three distinct hardware architectures for SPONGEWRAP-based authenticated encryption, secured against DPA. The chip is intended to be used as a fully flexible evaluation platform for determining the effectiveness of hiding and masking countermeasures in a real ASIC. Therefore, the three architectures solely differ with regard to the realized masking technique. While the first design makes use of the 3-share approach proposed by Bertoni et al. [2,7], the latter two utilize the 3-share implementation with re-masking and the 4-share approach presented by Bilgin et al. [8], which both fulfill the uniformity property. Moreover, each of the three architectures contains hiding countermeasures, which can be switched on and off at will. The main fields of application for ZORRO are resource-limited environments such as smart cards, embedded systems, or RFID-based devices, which is why low-area was our most important design goal. ZORRO was fabricated in a 180 nm CMOS process technology by UMC and the smallest of the three architectures requires only 14.5 kGE. This represents the smallest reported masked KECCAK ASIC implementation to date. Beside the un-keyed KECCAK implementations available in literature [4,14,20], the smallest reported masking-secured designs so far require more than 30 kGE [2,7,8].

---

[1] This mode involves a secret key that needs to be protected against implementation attacks. It is used in, e.g., stream encryption or authenticated encryption modes.

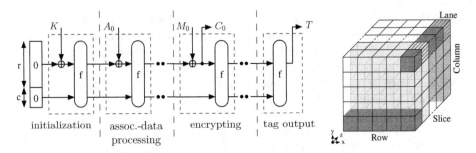

Fig. 1. The SPONGEWRAP construction.   **Fig. 2.** KECCAK state.

Moreover, we are the first to present DPA results targeting KECCAK implementations on a fabricated ASIC chip. We provide first DPA results of the unprotected, the hiding-secured, and the three share threshold implementation (TI). A higher-order DPA attack against the masked implementation succeeds with about 70 000 measurements. In order to reach a comparable security level with the hiding countermeasure, an impractical number of 240 dummy rounds needs to be inserted, what equals ten times the number of rounds (24) for one KECCAK-f permutation. Future work will consist in a detailed comparison of the three threshold implementations on the ASIC, requiring a huge amount of measurements which are not available yet.

The remainder of this paper is structured as follows. In Sect. 2, we give a brief introduction to the authenticated encryption mode SPONGEWRAP. Section 3 presents the hardware architecture of ZORRO. Implementation and power-analysis results are given in Sect. 4 and finally a discussion about the results and future work is provided in Sect. 5.

2   The SpongeWrap Construction

The core element of SPONGEWRAP is the KECCAK-f permutation [6]. The most prominent application of KECCAK-f is its use in a sponge construction [5] to build the hash algorithm KECCAK, which has recently been presented as a new draft for the upcoming SHA-3 standard by NIST [19]. However, KECCAK-f can also be used to form several other cryptographic primitives [3], including the AE mode SPONGEWRAP. In the following, a brief introduction to the SPONGEWRAP construction and the KECCAK-f permutation is given. For an in-depth discussion about the two primitives, we refer the reader to [3] and [6].

The SpongeWrap Construction. Figure 1 illustrates the SPONGEWRAP mode, which uses a duplex construction [3] to create an AE scheme. It can be subdivided into four phases: an *initialization* phase, an *associated-data processing* phase, an *encryption* phase, and a *tag-generation* phase. During the *initialization*, the state is cleared and loaded with the cipherkey K by a call to the permutation f. After that, the SPONGEWRAP object is able to receive data for *wrapping* associated data blocks A_i (authenticated only) and plaintext blocks M_j (authenticated and encrypted) to retrieve the ciphertext blocks C_j and the corresponding authentication tag T. The respective decryption process is known as *unwrapping* and basically swaps plaintext and ciphertext blocks and compares the received authentication tag with the recomputed one. If the two tags do not match, an error will be dumped, but no plaintext will be provided.

The Keccak-f Permutation. KECCAK-f operates on a state with a fixed size of b bits. This state consists of two parts: r (bitrate) and c (capacity), where r specifies the number of input bits, which are processed in one iteration and therefore relates to the speed of the computation. The last c bits of the state determine the attainable security level of the construction, i.e., $c = b - r$. The authors of KECCAK defined KECCAK-f for the following seven state sizes: $b = 25 \times w$, where $w = 2^\ell$ and ℓ ranges from 0 to 6. The state is organized as a $5 \times 5 \times w$ matrix with three dimension coordinates (x,y,z). We call a set of w bits with given (x,y) coordinates a *lane*, a set of 5 bits with given (y,z) coordinates a *row*, 5 bits with given (x,z) coordinates a *column*, and the 5×5 matrix for a given (z) coordinate a *slice* (see Fig. 2). The KECCAK-f function further consists of $12 + 2\ell$ rounds that are made up of five steps:

θ : Used to integrate diffusion by a linear mixing layer (the parity of two nearby columns is added to each column).

ρ : Inter-slice dispersion (all lanes are rotated by a defined offset).

π : Breaking horizontal/vertical alignment (the 25 lanes are transposed in a fixed pattern).

χ : The non-linearity part of KECCAK-f (the 5 bits of each row are combined using AND gates and inverters and the shifted result is added to the row).

ι : A w-bit round constant is added (XORed) to a single lane.

3   Hardware Architecture of Zorro

We intend to use ZORRO as an evaluation platform for investigating the quality of DPA countermeasures for an AE system based on the KECCAK-f[1600] permutation. Our main goal was to build an ASIC, providing different types of *masking* and *hiding* techniques. As pointed out by Bilgin et al. [8], a first-order DPA-secure KECCAK design, which is based on a three-share threshold implementation (without re-masking), does not fulfill the *uniformity property* [17,18] and thus, is not provable secure against first-order DPA attacks. Hence, we decided to place three distinct architectures on ZORRO, which only differ with regard to the implemented masking scheme. The first design is based on a three-share approach as proposed by Bertoni et al. [7]. The second and third architectures make use of the threshold implementation improvements presented by Bilgin et al. [8], namely a three-share design using re-masking and a four-share architecture. Figure 3 shows a block diagram of the top-level design entity, including the three distinct architectures named *3-Share*, *3-Share**, and *4-Share*.

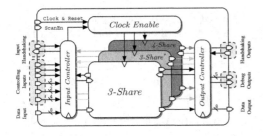

Fig. 3. Top-level architecture of ZORRO.

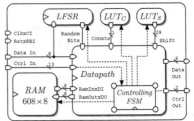

Fig. 4. Design of the *3-Share* entity.

In order to assure that meaningful power measurements can be taken from each distinct architecture separately, the *Clock Enable* entity contains clock gating cells, which enable the clock only for the actually selected entity. Moreover, the *Input Controller* forwards the input signals solely to the currently activated entity, thereby avoiding any logical changes in combinational paths within the deactivated architectures. With this setup, we are able to obtain meaningful power measurements of each design without significant noise from the deactivated units with regard to their dynamic power consumption. The *Output Controller* is responsible for forwarding the output signals of the respective unit once an input data block has been processed. Using a couple of debug outputs, ZORRO provides additional information about currently ongoing internal processes. Data to and from the chip can be transmitted via an eight bit data bus, controlled by a four way handshaking protocol. Each of the three architectures by itself can either operate in encryption or decryption and offers four different modes of operation:

Normal Mode: The normal mode represents the default mode in which no DPA countermeasures are activated. Hence, only one third (for the three-share based architectures) respectively one fourth (for the four-share based design) of the state-storing RAMs is actually used. Measurements based on this mode serve as a reference for the protected alternatives.

Hiding Mode: Running in this mode, ZORRO uses two hiding countermeasures in order to circumvent DPA attacks. First, the user can choose how many *dummy operations* should be executed during processing a single input block. Using a control signal, up to 15 dummy operations can be initiated, each of them representing a full round of the KECCAK-f[1600] permutation. Second, all three architectures can shuffle their computations by varying between eight different read/write addresses when accessing the RAM.

Masked Mode: When operating ZORRO in this mode, en-/decryption is performed using masking countermeasures in order to prevent DPA attacks.

Secure Masked Mode: In this mode, both hiding and masking countermeasures are activated and hence, this represents the most secure way on how to operate ZORRO with regard to its DPA security.

3.1   *3-Share*, *3-Share**, and *4-Share* Architectures

Since the *3-Share*, *3-Share**, and *4-Share* hardware architectures differ only very slightly, we will further on solely discuss the *3-Share* version and point out the differences to the other two architectures if necessary.

We aimed to design a low-area, DPA-secure authenticated encryption system based on KECCAK. Because of these goals, the area density of the memory required to store the KECCAK state is of utmost importance. Moreover, the implemented secret sharing countermeasure works on the algorithmic level, and thus the required memory for the state increases with each share. Therefore, we favored a random-access memory (RAM) macro cell over their standard cell counterparts to store the state, which offers a better bit-per-area density. We store both the round constants of the ι function and the shift offsets of the ρ function in look-up tables (LUTs). Figure 4 illustrates the uppermost hierarchy level of the *3-Share* entity, including the state RAM, the LUTs, and the datapath entity, which gets controlled by a finite-state machine (FSM). Moreover, Fig. 4 shows the linear-feedback shift register (LFSR), which is constructed by the primitive polynomial $x^{32} + x^7 + x^3 + x^2 + 1$. The initialization of the LFSR is done with an external seed. Its output is used on the one hand for determining whether to perform a dummy operation or not, and on the other hand for generating the random bits required for the re-masking in the *3-Share** architecture. Overall, 42 random bits are required per input block (39 for the dummy operation conditions and three for the shuffling of the RAM addresses).

The *3-Share* architecture contains a 608×8 RAM (cf. Fig. 4) for storing the state and the shares. Basically, a secret sharing scheme for KECCAK based on three shares would require only 4 800 bits (three times the state size). We use the additional eight bytes of the RAM as inputs for the dummy operations during the hiding mode and therefore, keep these memory locations uninitialized. Thereby, none of the dummy operations computes on the actual payload of the chip and hence, no correlated power figures should be observed.

For the initial masking of the *3-Share* (*4-Share*) entity, the chip receives 3 200 (4 800) random bits to initialize two (three) shares followed by the plaintext. The last share equals the XOR-sum of the already initialized shares with the plaintext. The implementation of the KECCAK-$f[1600]$ permutation is based on a combined lane and slice processing, similar to that proposed by Pessl and Hutter [20]. Figure 5 shows the architecture of the *Datapath* unit of the *3-Share* entity. We use the *SubState* register to buffer lanes and slices currently being processed.

RAM Allocation. As proposed by Bertoni et al. [4], storing the bits of lanes and slices in an interleaved form allows efficient processing of the data when choosing a small datapath width, meanwhile keeping the size of the required buffer register at a minimum. We also make use of this technique and store four bits of two slices in each RAM word (i.e., two bits of four lanes). Since we need four lanes at a time, this results in a buffer register of 256 bits. Unfortunately, the state consists of 25 lanes and thus, not all lanes can be stored in this interleaved form. We decided to store the first lane in a linear way, as this lane is not influenced by the

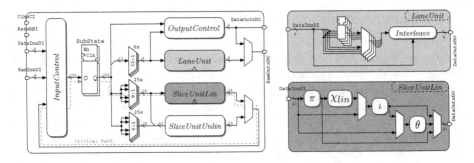

Fig. 5. Datapath of the *3-Share* entity (controlling signals omitted).

ρ operation and hence, can be skipped for this function. Although, based on this memory allocation we waste a negligible amount of clock cycles when loading data of the first lane, we can keep the size of the *SubState* register comparatively small. In order to avoid switching back and forth between slice-based and lane-based operations as much as possible, we make use of the same rescheduling approach proposed in [20], where they distinguish between the following three different types of "rounds":

$$R_1 = \theta \times \rho \qquad R_{2...24} = \pi \times \chi \times \iota \times \theta \times \rho \qquad R_{25} = \pi \times \chi \times \iota \qquad (1)$$

Round Operations. When ZORRO operates in normal mode, the four slice-based round functions of KECCAK-f (θ, π, χ, and ι) are exclusively calculated in the *SliceUnitLin* within a single clock cycle for a whole slice. The applied round schedule requires to calculate the result of θ, $\pi \times \chi \times \iota \times \theta$, and $\pi \times \chi \times \iota$. As illustrated on the bottom-right of Fig. 5, all three operations can be accomplished within the *SliceUnitLin* with the use of bypass multiplexers. Calculations of the linear round functions of the KECCAK-f permutation are equal for both the normal mode and the masking-secured modes. Here, each share can be computed in sequential order (e.g., in R_1 the theta step is performed three (four) times sequentially in order to process the three (four) shares). Due to the fact that the non-linear χ function requires inputs from more than one share, the processing of this function slightly differs. For the hardware implementation of the *3-Share* architecture, we follow the approach presented by Bertoni et al. [7] and compute the result for two input slices in a single cycle within the *SliceUnitUnlin* entity. For the lane-based operation (ρ), we aimed at calculating its output byte-by-byte. This allows us to combine it with the RAM write operation. Thanks to the chosen RAM allocation, multiples of two-bit-wide shift operations of lanes can easily be accomplished with the addressing of the memory. The special storage structure provides information about four lanes per RAM word (byte) and the *SubState* register can hold up to four lanes simultaneously. Unfortunately, each lane has a different shift offset. Hence, different bit couples of the buffered lanes must be taken to compensate the differences between the offsets. The different compensation offsets can be precalculated and are stored in the LUT_S entity

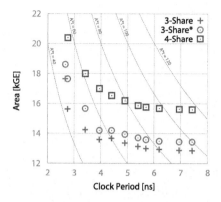

Fig. 6. AT plot of ZORRO's three different architectures obtained after synthesis.

Table 1. Area breakdown of the ZORRO ASIC (synthesis results at 5 ns).

Component	Area [GE]	Area [%]
3-Share	13 370	30.5
Datapath &FSM	7 300	16.7
RAM	4 680	10.7
LFSR	300	0.7
SliceUnitLin	480	1.1
Others	610	1.3
*3-Share**	13 940	31.8
4-Share	16 190	37.0
I/O Interface	320	0.7
Zorro Total	43 820	100.0

(see Fig. 4) for each lane quadruple. With these precalculated values, multiples of two-bit-wide shift operations, and the offset between the different lanes can be compensated. The leftover is a possible shift by one bit. Therefore, 4 one-bit-registers with surrounding multiplexers are used. If a lane is shifted by one bit, the high bit of the chosen bit couple is stored in the one bit register. The low bit is shifted one bit to the left and the old content of the one bit register is used as the new low bit. This is done for each bit couple of the buffered lanes. The result is stored back to the RAM in interleaved form. The responsible unit for the lane-based operation is called *LaneUnit* (cf. Fig. 5).

4   Results

The results of our work are twofold. First, we present our implementation results of ZORRO and provide actual ASIC performance numbers of the *3-Share*, *3-Share**, and *4-Share* design. Second, we present first practical results of DPA investigations on our AE system using power traces obtained from the real chip.

4.1   Hardware Figures of Zorro

We used VHDL in order to code the RTL model of ZORRO and Mentor Graphics' ModelSim version 10.2a to verify its functional correctness. Synthesis results were obtained from Synopsys' Design Compiler version 2012.06 for a mature 180 nm CMOS technology by UMC. The designs were synthesized using a standard cell library by Faraday Technologies under typical case conditions and backend design steps were accomplished using SoC Encounter from Cadence. Area results will be given in terms of gate equivalents (GEs), for which one GE equals the size of a two-input NAND gate of the utilized standard cell library (= 9.3744 µm²).

In order to provide a fair comparison between the results of ZORRO and related work as well as meaningful numbers for an actual chip to be taped out,

Table 2. Comparison of ZORRO with related ASIC designs (synthesis results).

Source	Techn. [nm]	Area [GE]	f_{max} [MHz]	Perf.[a] [Cycles]
Designs w/o DPA Countermeasures				
Pessl and Hutter [20][b]	130	5 522	61	22 570
Bilgin et al. [8][c]	180	10 800	555	1 600
ZORRO in Normal Mode[b]	180	13 370	200	21 888
3-Share-Secured Designs w/o Re-Masking				
Bertoni et al. [7][c]	130	95 000	200	72
ZORRO *3-Share* Architecture[b]	180	13 370	200	113 184
3-Share-Secured Designs w/ Re-Masking				
Bilgin et al. [8][c]	180	33 100	553	1 625
ZORRO *3-Share*[*] Architecture[b]	180	13 940	200	113 184
4-Share-Secured Designs				
Bilgin et al. [8][c]	180	43 100	572	1 600
ZORRO *4-Share* Architecture[b]	180	16 190	200	149 640

[a] KECCAK-f permutation
[b] Block size of 1088 bits
[c] Block size of 1024 bits

we present two different area numbers. On the one hand, we provide synthesis results without considering any Design for Testability (DFT) techniques.[2] On the other hand, we include the area numbers after all backend design steps have been successfully accomplished and therefore the designs include DFT circuitries for RAM tests as well as scan flip-flops to enable automated test pattern generation (ATPG). Figure 6 provides an area/time (AT) plot of the synthesis results of the three different architectures. Based on the isolines, indicating a constant AT product, it can be observed that for a clock period below 4 ns, the resulting area of each architecture increases significantly. Moreover, we decided to spend some room for the upcoming backend run and therefore, chose a maximum frequency of 200 MHz for ZORRO. The critical path of the design runs through the *SliceUnitLin* entity, highlighted using a dashed line in Fig. 5. From Fig. 6 it can be seen that the area differences between the three architectures remains quite constant. This was expected since a major part of the overall area is occupied by the RAM. Other differences between the three designs with regard to their logic components are almost negligible. Table 1 lists an area breakdown of the ZORRO ASIC after synthesis for 5 ns. It shows that our *3-Share*, *3-Share*[*], and *4-Share* architectures require 13.4 kGE, 13.9 kGE, and 16.2 kGE, respectively. Table 2 lists a comparison between ZORRO and related KECCAK-based ASIC designs in the field of low-area and DPA-security.

---

[2] Note that such numbers can vary significantly compared to the actual area figures of a finalized chip ready for tapeout, depending on the implemented design.

Fig. 7. Chip layout and photo of ZORRO.

Table 3. Datasheet of ZORRO.

Property	
Technology (UMC)	0.180 µm
Supply Volt. (Core/Pad)	1.8 V/3.3 V
Max. Frequency (f_{max})	200 MHz
Required Area	46.0 kGE
Est. Power Cons. @ f_{max}	
3-Share	17.3 mW
*3-Share**	19.7 mW
4-Share	20.8 mW
Crypt. Perf. (Normal/Masked)[a]	
3-Share	21 888/113 184
*3-Share**	21 888/113 184
4-Share	21 888/149 640

[a] Requ. cycles for one KECCAK-f perm.

For the actual tapeout-version of ZORRO, we added a couple of DFT circuitries in order to provide suitable testing possibilities. This, the insertion of the required buffers, and the fact that after the backend design a realistic wire-load model was available, lead to an increase in area to 14 kGE, 14.5 kGE, and 17 kGE for the *3-Share*, *3-Share**, and *4-Share* architectures, respectively.

Figure 7 shows the final layout of ZORRO as well as a photo of the chip. Table 3 provides a datasheet for some of ZORRO's final specifications.

4.2 Power-Analysis Results

In order to validate our design regarding power-analysis resistance, we performed power measurements and applied a standard Correlation Power Analysis (CPA) based on the Pearson correlation coefficient [10] on the measured power traces as a first step. Furthermore, higher-order CPA attacks were performed targeting the *3-Share* and *3-Share** implementations. For the rest of this section, ρ_c indicates the correlation coefficient of the correct key guess. Another procedure to rate the power-analysis resistance of an implementation is the method presented in [13] based on the statistical t-test. The advantage of the t-test is that no leakage model has to be defined. We have observed Hamming-distance leakage of intermediate values during simulation runs, so we decided to perform CPA attacks as a first step. For future work we will also investigate the t-test methodology and compare the outcome with the results presented in this work. Due to the time-consuming measurement process and the huge number of required measurements, we did not investigate the *4-Share* implementation so far.

We used a *Picoscope 6404c* oscilloscope to capture the power traces from the ZORRO ASIC. The voltage drop across a 1 Ω resistor in the core supply line was measured by applying a *LeCroy AP033* differential probe. This setup allows to minimize the noise created by, for instance, I/O activity of the chip because the chip has a separate supply line for the I/O part. The traces were recorded with a sampling rate of 1 GS/s and the clock frequency of the ASIC was set to 10 MHz.

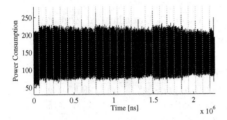 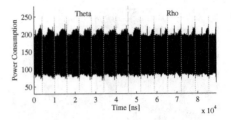

Fig. 8. Power trace of an entire KECCAK-f permutation while ZORRO is running in *normal mode* (left plot). Zoom into the first round, computing θ and ρ (right plot).

The First Power Trace. The left plot in Fig. 8 shows a measured power trace of an entire KECCAK-f permutation of ZORRO running in *normal mode*. It shows all 24 rounds (including one additional round at the end where ρ is skipped) separated by a dotted vertical line. The right plot in Fig. 8 shows a zoom into the first round. We separated the slice and lane processing phases with a dashed vertical line as well as the eight slice-processing iterations (ZORRO processes the 64 slices in eight blocks) by dotted vertical lines. The same was done for the six lane-processing iterations (ZORRO processes all 24 lanes in blocks of four). The time interval where the θ step of the first round takes place is of special interest because the power-analysis attacks presented in the following target the θ step. Only the first θ step was recorded for the power analysis attacks in order to keep the amount of data small.

Performing CPA. CPA attacks presented in this work focus on the first round of KECCAK-f. In particular, we targeted a storage operation of the 256-bit *Sub-State* register that stores key-dependent intermediate values during the θ step. The decision to target the θ transformation and not the non-linear χ transformation was motivated by the modified round schedule. In the first round, θ is the only slice-based transformation leading to a simple power model. We target a (unknown but constant) 256-bit key that gets concatenated with the (known and random) associated data. Thus, each targeted slice operation reveals information about four key bits. Since θ processes two slices in parallel, we can efficiently target 8 key bits by evaluating 256 key hypotheses. θ is a linear function, so not only the correct key will result in a high correlation, but shifted key variants will correlate too. Therefore, this attack does not only reveal the correct key but it will also reveal a small set of other possible key candidates (in our experiment we will get up to eight out of 256 possible key candidates). Due to the fact that we can attack all 64 slices and four bits of the key of two subsequent slice pairs must be similar, only eight key candidates with a length of 256 bits remain for a brute-force attack, what is within computational bounds (e.g., when attacking the slice pair (1, 2) followed by the slice pair (2, 3), only key candidates where the key bits of slice two are similar need to be considered). First experiments showed that ZORRO leaks the information according to the Hamming distance power model [16]. As a reference, preliminary attacks target ZORRO running in

Table 4. Results for the power-analysis attacks on ZORRO running in *hiding mode (HM)*.

Mode	ti	Windowing	$\rho_{c,theory}$	$\rho_{c,pract}$
HM 1	16	no	0.044	0.049
HM 1	16	yes	0.176	0.237
HM 2	24	no	0.029	0.031
HM 2	24	yes	0.152	0.160
HM 15	128	no	0.005	-
HM 15	128	yes	0.062	0.057

Table 5. Min. number of measurements required.

Mode	N_{meas}
NM	< 100
HM 1	285
HM 2	625
HM 15	4 925
HM 240[a]	70 000
MM 3-share	70 000

[a] not supported by ZORRO

normal mode (NM, no countermeasures enabled). After the initial attacks, we have activated the DPA countermeasures one after the other in order to evaluate their impact on the power-analysis attacks. We first evaluate the *hiding mode (HM)* followed by the *masked mode (MM)*.

Normal Mode: The CPA attack was performed with 1 000 power traces leading to $\rho_c = 0.7$[3]. This ρ_c value indicates that less than 50 measurements are sufficient to distinguish the correct key hypothesis from the wrong key hypotheses [16].

Hiding Mode: Next, we have activated hiding on the ZORRO ASIC. The *number of dummy rounds* (N_{dr}) has been set to 1 *(HM 1)*, which means that zero or one dummy operation is randomly inserted in front of the first KECCAK-f permutation working on the real data. Moreover, as soon as the hiding mode is activated, the execution order during each slice-processing operation is randomized. As a result, the targeted operation can appear at 16 different *time instances ti*. According to [16], ρ_c should decrease by a factor of $\frac{1}{ti}$ compared to the unprotected case. When taking into account ρ_c of 0.7 for the unprotected case and $ti = 16$, this leads to an expected $\rho_{c,theory}$ value for the protected implementation of $\frac{0.7}{16} = 0.044$. Attacks on the protected implementation yield $\rho_{c,pract} = 0.049$, what fits well with theory. Next, windowing has been applied combining all the 16 time instances. According to [16], windowing should increase ρ_c by a factor of $\sqrt{16}$ for our attack. With windowing applied, our practical attacks yield $\rho_{c,pract} = 0.237$, what is significantly higher than the expected value $\rho_{c,theory} = 0.176$. Further practical experiments have been performed with $N_{dr} = 2$ *(HM 2)* leading to a $\rho_{c,pract}$ value of 0.031 without and 0.160 with windowing applied. Again, these values fit well with theory. ZORRO allows a maximum N_{dr} of 15 *(HM 15)*, here practical results with windowing applied yield $\rho_{c,pract} = 0.057$ what is again close to $\rho_{c,theory} = 0.062$. Without windowing applied, no significant results can be observed with 100 000 measurements. Due to that reason, for *HM 15* without windowing we can only give $\rho_{c,theory} = 0.005$. Table 4 summarizes the results with regard to the hiding mode.

---

[3] The confidence interval of the coefficient, where 99.99 % of all samples (4-σ border) are located in the normal distribution model for 1 000 traces, is about 0.12.

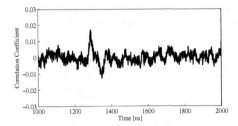

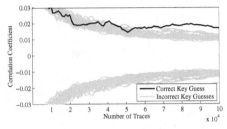

Fig. 9. 3rd-order CPA result for the correct key guess using 100 000 ASIC traces (ZORRO running in *masked mode*).

Fig. 10. Course of the correlation coefficient of ZORRO running in *masked mode* (3rd-order CPA).

Masked mode: In a next experiment we performed power-analysis attacks targeting the first θ step on ZORRO running in *masked mode* (hiding was deactivated for this experiment). 1st-order CPA attacks using 100 000 power traces captured from the ASIC did not succeed. No significant correlation peaks could be observed in the result. Due to the clear patterns in the power traces, the time instances, where the first θ steps of each share are performed, can be identified with small effort. By combining the revealed time instances, a 3rd-order CPA attack has been mounted. The centralized product combining has been used as combination function, as suggested by Prouff et al. [21]. As shown in Fig. 9, this attack results in a significant correlation peak for the correct key hypothesis with $\rho_c = 0.016$. Figure 10 shows the course of the correlation coefficients for all key guesses. With less than 70 000 measurements the correct key hypothesis can be distinguished from the wrong key hypotheses. Note that since the modifications between the *3-Share* and *3-Share** implementation solely affect the χ step (and not the herein targeted θ operation), the results of the 3rd-order CPA are identical for both three-share based architectures.

Comparison masking mode and hiding mode: Our first attack results show that both hiding as well as masking increase the effort for an attack. Attacks on the implementation using hiding also succeed without any modification of the traces (e.g., windowing). But, if windowing is applied, ρ_c only decreases by a factor of $\frac{1}{\sqrt{ti}}$ instead of $\frac{1}{ti}$. That means, windowing drastically reduces the number of required measurements N_{meas} for performing a successful attack. Attacks on the masked implementation do not succeed without combination of the traces, at least not if the shares are calculated sequentially during the first θ step, as it is the case with ZORRO. In order to reach the same security level with hiding as with masking, 240 dummy rounds would be required. Each additional dummy round leads to eight additional time instances for the targeted operation to appear, so $ti = 240 \cdot 8 = 1920$ for 240 dummy rounds. As a consequence, ρ_c decreases to $\frac{0.7}{\sqrt{1920}} = 0.016$. This is now equal to ρ_c achieved with a 3rd-order CPA targeting the masked implementation. However, with 240 dummy rounds, the runtime of the implementation running in *hiding mode* exceeds the runtime of the implementation running in *masked mode*. Table 5 summarizes the number of required measurements N_{meas} for a successful key recovery. For the unprotected case (NM), less than 100 measurements are sufficient, for the hiding mode with the weakest protection $(HM\,1)$, $N_{meas} = 285$, and for the hiding mode with the highest protection $(HM\,15)$, $N_{meas} = 4\,925$ respectively. Note that N_{meas} for $HM\,1$, $HM\,2$, and $HM\,15$ all assume that windowing is applied. Successful attacks targeting the masked mode $(MM\,3\text{-}share)$ require 70 000 measurements. 240 dummy rounds $(HM\,240)$ would also yield $N_{meas} = 70\,000$ but this mode is not supported by ZORRO since 15 is the maximum number of dummy rounds.

5   Conclusions and Future Work

ZORRO represents the first actually taped-out ASIC, hosting KECCAK-based authenticated encryption systems secured against DPA attacks. It contains three distinct architectures, which solely differ with regard to the implemented secret-sharing technique. In addition to the DPA-secure designs, we aimed at low-area hardware architectures, targeting resource-constrained applications.

As future work, we are looking forward to investigate more DPA attacks targeting different intermediates of KECCAK. In addition to target the output of the θ step, we want to target the output of the χ step as effects like glitches or early-propagation might allow successful DPA attacks with a lower order than 3. This will also enable a comparison of the three masking-secured implementations. Furthermore, we want to increase the number of traces up to several millions to identify unintended leaks. Finally, we plan to apply powerful Test Vector Leakage Assessment (TVLA) tests (including the fixed-vs-random t-test) to detect non-specific leakages and to verify the DPA resistance of our cores.

Acknowledgements. This work has been supported in part by the Swiss Commission for Technology and Innovation (CTI) under project number 13044.1 PFES-ES and in part by the European Commission through the FP7 program under project number 610436 (project MATTHEW) and by the Austrian Research Promotion Agency (FFG) and the Styrian Business Promotion Agency (SFG) under grant number 836628 (SeCoS). Moreover, we would like to thank the people from the Microelectronics Design Center of ETH Zurich for their support during the backend design of the ASIC. We also want to thank Svetla Nikova and Joan Daemen for their comments on the pre-print version of the paper.

References

1. CAESAR: Competition for Authenticated Encryption: Security, Applicability, and Robustness, March 2013. http://competitions.cr.yp.to/caesar.html
2. Bertoni, G., Daemen, J., Debande, N., Le, T.-H., Peeters, M., Van Assche, G.: Power analysis of hardware implementations protected with secret sharing. Cryptology ePrint Archive: Report 2013/067, February 2013
3. Bertoni, G., Daemen, J., Peeters, M., Van Assche, G.: Duplexing the sponge: single-pass authenticated encryption and other applications. In: Miri, A., Vaudenay, S. (eds.) SAC 2011. LNCS, vol. 7118, pp. 320–337. Springer, Heidelberg (2012)

4. Bertoni, G., Daemen, J., Peeters, M., Assche, G.V., Keer, R.V.: Keccak Implementation Overview, May 2012. http://keccak.noekeon.org/Keccak-implementation-3.2.pdf (Version 3.2)
5. Bertoni, G., Daemen, J., Peeters, M., Van Assche, G.: Sponge functions. In: ECRYPT Hash Workshop, Barcelona, Spain, 24–25 May 2007. http://sponge.noekeon.org/SpongeFunctions.pdf
6. G. Bertoni, J. Daemen, M. Peeters, and G. Van Assche. KECCAK specifications, Version 2, 10 September 2009. http://keccak.noekeon.org/Keccak-specifications-2.pdf
7. Bertoni, G., Daemen, J., Peeters, M., Van Assche, G.: Building power analysis resistant implementations of keccak. In: 2nd SHA-3 Candidate Conference (2010)
8. Bilgin, B., Daemen, J., Nikov, V., Nikova, S., Rijmen, V., Van Assche, G.: Efficient and first-order DPA Resistant implementations of keccak. In: Francillon, A., Rohatgi, P. (eds.) CARDIS 2013. LNCS, vol. 8419, pp. 187–199. Springer, Heidelberg (2014)
9. Borisov, N., Goldberg, I., Wagner, D.: Intercepting mobile communications: the insecurity of 802.11. In: Naghshineh, M., Zorzi, M., (eds.) MobiCom 2001, pp. 180–189. ACM (2001)
10. Brier, E., Clavier, C., Olivier, F.: Correlation power analysis with a leakage model. In: Joye, M., Quisquater, J.-J. (eds.) CHES 2004. LNCS, vol. 3156, pp. 16–29. Springer, Heidelberg (2004)
11. Canvel, B., Hiltgen, A.P., Vaudenay, S., Vuagnoux, M.: Password interception in a SSL/TLS channel. In: Boneh, D. (ed.) CRYPTO 2003. LNCS, vol. 2729, pp. 583–599. Springer, Heidelberg (2003)
12. Gérard, B., Grosso, V., Naya-Plasencia, M., Standaert, F.-X.: Block ciphers that are easier to mask: how far can we go? In: Bertoni, G., Coron, J.-S. (eds.) CHES 2013. LNCS, vol. 8086, pp. 383–399. Springer, Heidelberg (2013)
13. Gilbert Goodwill, B.J., Jaffe, J., Rohatgi, P.: A testing methodology for side-channel resistance validation. In: NIST Non-invasive attack testing workshop (2011)
14. Kavun, E.B., Yalcin, T.: A lightweight implementation of keccak hash function for radio-frequency identification applications. In: Ors Yalcin, S.B. (ed.) RFIDSec 2010. LNCS, vol. 6370, pp. 258–269. Springer, Heidelberg (2010)
15. Kocher, P.C., Jaffe, J., Jun, B.: Differential power analysis. In: Wiener, M. (ed.) CRYPTO 1999. LNCS, vol. 1666, pp. 388–397. Springer, Heidelberg (1999)
16. Mangard, S., Oswald, E., Popp, T.: Power Analysis Attacks - Revealing the Secrets of Smart Cards. Springer, New York (2007). ISBN 978-0-387-30857-9
17. Nikova, S., Rijmen, V., Schläffer, M.: Secure hardware implementation of non-linear functions in the presence of glitches. In: Lee, P.J., Cheon, J.H. (eds.) ICISC 2008. LNCS, vol. 5461, pp. 218–234. Springer, Heidelberg (2009)
18. Nikova, S., Rijmen, V., Schläffer, M.: Secure hardware implementation of nonlinear functions in the presence of glitches. J. Cryptology $24(2)$, 292–321 (2011)
19. NIST. SHA-3 Standard: Permutation-Based Hash and Extendable-Output Functions (DRAFT FIPS PUB 202), May 2014
20. Pessl, P., Hutter, M.: Pushing the limits of SHA-3 hardware implementations to fit on RFID. In: Bertoni, G., Coron, J.-S. (eds.) CHES 2013. LNCS, vol. 8086, pp. 126–141. Springer, Heidelberg (2013)
21. Prouff, E., Rivain, M., Bévan, R.: Statistical analysis of second order differential power analysis. IEEE Trans. Comput. $58(6)$, 799–811 (2009)

Hands-on Side-Channel Analysis

Side-Channel Security Analysis of Ultra-Low-Power FRAM-Based MCUs

Amir Moradi[✉] and Gesine Hinterwälder

Horst Görtz Institute for IT-Security, Ruhr-Universität Bochum, Bochum, Germany
{amir.moradi,gesine.hinterwaelder}@rub.de

Abstract. By shrinking the technology and reducing the energy requirements of integrated circuits, producing ultra-low-power devices has practically become possible. Texas Instruments as a pioneer in developing
FRAM-based products announced a couple of different microcontroller
(MCU) families based on the low-power and fast Ferroelectric RAM technology. Such MCUs come with embedded cryptographic module(s) as
well as the assertion that – due to the underlying ultra-low-power technology – mounting successful side-channel analysis (SCA) attacks has
become very difficult. In this work we practically evaluate this claimed
hardness by means of state-of-the-art power analysis attacks. The leakage sources and corresponding attacks are presented in order to give
an overview on the potential risks of making use of such platforms in
security-related applications. In short, we partially confirm the given
assertion. Some modules, e.g., the embedded cryptographic accelerator,
can still be attacked but with slightly immoderate effort. On the contrary,
the other leakage sources are easily exploitable leading to straightforward
attacks being able to recover the secrets.

1 Introduction

Side-Channel Analysis (SCA) attacks have become a serious threat to cryptographic implementations. Regardless of the theoretical robustness of a cryptographic primitive, secret materials used by its implementation can easily be
recovered in case of absence of SCA-dedicated countermeasures. Case studies
like [1,4,11–13] confirmed the effectiveness of such attacks to overcome the security of commercial products. Hence, the producers of security-related applications have moved towards integrating SCA countermeasures. For example the
FPGA architecture *UltraScale* [18] – recently announced by Xilinx – offers many
security features including DPA-protected bitstream encryption. Along the same
lines, Microsemi has integrated many solutions to improve physical security of
Actel's FPGA family *SmartFusion2* [10].

Texas Instruments (TI) has introduced ultra-low power FRAM-based microcontrollers (MCUs) with a couple of security features [15,16]. Ferroelectric RAM
(FRAM) technology enables large-scale non-volatile memories that offer faster
write operations, much larger tolerated number of write cycles, and a much lower

© Springer International Publishing Switzerland 2015
S. Mangard and A.Y. Poschmann (Eds.): COSADE 2015, LNCS 9064, pp. 239–254, 2015.
DOI: 10.1007/978-3-319-21476-4_16

power consumption compared to equivalent flash memories. The low power consumption as well as the embedded cryptographic modules (e.g., an AES core) are the key factors of the offered security. It is claimed that due to the ultra-low-power feature of such MCUs and their low operating voltage (1.5 V) SCA attacks become extremely difficult to mount.

This article deals with the aforementioned features and presents practical investigation results with respect to the claimed hardnesses. The results of our analyses on an MSP430FR5969 MCU are summarized as:

- A couple of different power analysis attacks are feasible on the embedded AES module. We should highlight that such attacks are not as straightforward as those mounted on crypto engines of other MCUs e.g., Atmel's XMEGA [8]. That difference is mainly due to the low-power feature of the integrated AES module.
- Regardless of the underlying low-power architecture, software implementations of cryptographic algorithms executed on the underlying MCU are victims of power analysis attacks. Unsurprisingly, the secrets of such implementations can be easily revealed by means of straightforward state-of-the-art attacks.
- Due to the restricted speed of the FRAM technology, TI integrated a dedicated cache to be used when the MCU operates at a higher frequency than the access frequency of FRAM. As a known issue, the cache (miss/hit) can be a source of SCA leakage. We report case studies, which make use of this feature to launch effective SCA attacks.
- The internal architecture of MCUs is usually not known to end users. Such architectures can turn an implementation of a sound masking scheme to a vulnerable design. In order to examine such issues we consider an implementation of the *masking with randomized look-up table* countermeasure [14] which has particularly been developed for FRAM-based MCUs [7]. Our analysis shows that the unknown internal architecture of the underlying MCU causes the provably-secure masked design to have a first-order leakage, while it is supposed to provide security at all orders.

2   Features

Here we shortly recall a couple of features of TI's FRAM-based MCUs. We focus on those specifications, which are related to our security analyses.

2.1   AES Accelerator

In many of TI's FRAM-based MCUs – including the MSP430FR59xx family – an AES accelerator module is embedded. It supports both encryption and decryption for all key lengths (128, 192, and 256). Further, on-the-fly as well as offline round key generation scenarios are supported, and it is facilitated to be used in ECB, CBC, OFB, and CFB modes of operation. It should be highlighted that the AES module has not been designed for speed-critical applications although

Table 1. AES accelerator performance figures

Key length	Encryption (clock cycles)	Decryption (clock cycles)
128 bits	168	168
192 bits	204	206
256 bits	234	234

it can perform a complete encryption and decryption much faster than corresponding software on the same MCU. Table 1 shows the number of clock cycles the AES module requires to complete the respective operations.

As it is a stand-alone module, the MCU can perform other operations while the AES module is busy. It is noteworthy that the numbers given in this table are with respect to the on-the-fly computation of round keys while the decryption module requires to receive the last round key. The interested reader is referred to [17] for more detailed information including the performance of the other modes.

2.2 FRAM Architecture

As a promising alternative to non-volatile storage such as flash, FRAM technology offers many advantages. It avoids the long delays as well as the high current supply required for programming (writing). The advantageous features of the FRAM technology focus mainly on write operations. High speed (125 ns delay), low power ($82\,\mu A/MHz$), and super high (10^{15}) write cycles have been reported for the 130 nm MSP430FR family of TI's MCUs.

As a disadvantage we should refer to the fact that FRAM reads are destructive. That is, every read must be followed by a write operation (with the same data). However, this is automatically handled by the FRAM controller, and the end user does not need to pay any attention to this. Therefore, the frequency of FRAM read operations are limited to the write speed. Due to this limit, TI has integrated a read cache in front of the FRAM to accelerate the operations in case the MCU operates at a higher frequency than the FRAM. In the MSP430FR family, the FRAM can be operated at up to 8 MHz without use of this cache. When the MCU operates at a frequency of 16 MHz (the maximum operation frequency of the MSP430FR family), the cache is utilized.

The integrated cache is a two-way associative cache containing two cache sets [17]. Each of these sets consists of two lines of four words (64 bits). The cache controller selects one of the cache lines to preload FRAM data and preserves recently-accessed data in the other cache line. If one of the four words stored in one of the cache lines is requested (a cache hit), no FRAM access occurs, and the requested data is read from the cache with full system speed. However, if none of the words that are available in the cache is requested (a cache miss), a wait state (one clock cycle at 16 MHz) controls the CPU to ensure proper

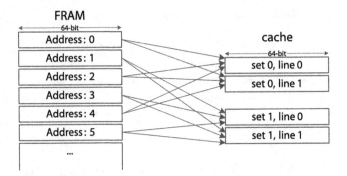

Fig. 1. Structure of the two-way associative cache

FRAM access. Therefore, memory read accesses on consecutive addresses can be executed without wait states when they are within the same cache line.

Each 64-bit location in FRAM can be cached in only one of the two sets in the cache. As shown in Fig. 1, the most common scheme is to use the least significant bit of the FRAM location's address as the indicator to the corresponding cache set. We should emphasize that FRAM contains both program code and data, e.g., look-up tables, which are to be stored in the non-volatile memory. Hence, frequent jumps and frequent accesses to the pre-stored tables in FRAM can negatively affect the cache performance.

3   Analyses

In this section we present various analyses that we performed on an FRAM-based MCU. We first present the framework that we used, and then describe each analysis in detail.

3.1   Setup

The practical analyses have been conducted on an MSP-EXP430FR5969 Launch-pad Evaluation Kit, and we used IAR Embedded Workbench IDE as well as Code Composer Studio to develop and compile the codes. This evaluation platform has been developed to facilitate power measurements. As shown in Fig. 2, we could easily place a $1.8\,\Omega$ resistor at the VCC path of the MSP430FR5969 MCU while no stabilizing capacitor was placed between the measurement point and the MCU. We monitored the current passing through the MCU by means of a LeCroy WaveRunner HRO 66Zi digital oscilloscope at a sampling rate of $1\,\mathrm{GS/s}$. We also used an I/O pin of the MSP-EXP430FR5969 Launchpad Evaluation Kit as trigger signal to align the collected traces. We provided a 16 MHz crystal oscillator as external clock source, and by clock source configurations drove the MCU at our desired frequency (explained below for each target).

Due to the very low power consumption of the MCU, we employed a DC blocker (BLK-89-S+ from Mini-Circuits) and an AC amplifier (ZFL-1000LN+

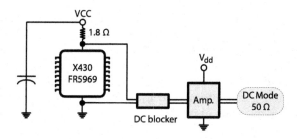

Fig. 2. Measurement setup

from Mini-Circuits) to collect the power traces with a considerably high quality. Further, we limited the oscilloscope bandwidth to 20 MHz to reduce the electrical noise.

Metrics: For the side-channel analysis we mainly used correlation power analysis (CPA) [3] to mount key-recovery attacks. However, in some cases we applied a statistical t-test [5]. The goal of such a scheme is not to examine a key-recovery attack, but rather it provides an overview of the existence of a leakage which might be exploited by an attack. The concept of t-test is based on the classical DPA attack of [9].

Following the concept of DPA, the traces $t \in T$ are categorized into two groups g_1 and g_2. Recall that Welch's (two-tailed) t-test is computed as

$$t = \frac{\mu(t \in g_1) - \mu(t \in g_2)}{\sqrt{\frac{\delta^2(t \in g_1)}{|g_1|} + \frac{\delta^2(t \in g_2)}{|g_2|}}},$$

where μ and δ^2 denote the sample mean and the sample variance respectively, and $|.|$ stands for the cardinality. The t-test indeed examines the validity of the z as the samples in both groups (g_1 and g_2) were drawn from the same population. If the null hypothesis is correct, it can be concluded with a high level of confidence that a corresponding DPA attack cannot exploit the leakage.

For such a conclusion the Student's t-distribution density function in addition to the degree of freedom is applied to determine the probability of rejecting the aforementioned hypothesis (for more information see [5]). For typical evaluations, a threshold for $|t|$ as > 4.5 is defined to reject the null hypothesis and conclude that the device exhibits a first-order leakage. This process is repeated at each sample point independently.

The remaining point to mention is the way that the categorization of traces into the groups g_1 and g_2 is performed. For a *specific* t-test this classification is done based on a chosen intermediate value. During the measurements the input (plaintext) is taken randomly while the key is kept constant for all the collected traces. With respect to the corresponding DPA attack – for example – an Sbox output bit directs the classification. Many intermediate values should be considered in the evaluations to examine the feasibility of each corresponding DPA attack. Instead, a *non-specific* t-test can be performed, which can examine the existence of leakage without any required intermediate value. In such a test, a fixed input (plaintext) is selected, and the measurements are randomly interleaved between the fixed and random inputs. So the non-specific t-test is also called *fixed vs. random* t-test. Hence, based on the given input (fixed or random) the traces are categorized into g_1 and g_2. It is noteworthy that such a leakage assessment scheme has been also used in [2].

3.2   AES Hardware Accelerator

As the first target we focus on the AES accelerator. As stated in Sect. 2.1, the AES module can perform the encryption and decryption functions in a couple of different settings. We evaluate only the AES-128 encryption function with on-the-fly round key computation, which takes 168 clock cycles. Following the configurations given in [17] we developed assembly code (in IAR Embedded Workbench IDE) to activate the aforementioned function. We intentionally wrote the code in such a way that it waits in a loop till the operation of the requested encryption is finished. It allows us to observe only the leakage of the AES accelerator module. Further, we configured the MCU to operate at a frequency of 2 MHz.

Figure 3(a) shows an exemplary power trace confirming its ultra-low power consumption. In order to examine the vulnerability of such a module to power analysis attacks, we first performed a couple of specific t-tests with intermediate values including i) the cipher round output bits (128 cases), ii) XOR between the cipher input and output bits (128 cases), iii) the SubBytes output bits (128 cases), iv) XOR between the SubBytes input and output bits (128 cases), v) the SubBytes output bytes (16×256 cases), and vi) XOR between different Sbox output bits ($\binom{16}{2} \times 8$ cases). The best results have been achieved considering the SubBytes output bits as well as the XOR between the SubBytes input and output bits (cases iii and iv). As a proof of concept we performed CPA attacks with the corresponding power models. For instance, Fig. 3 presents the results of two CPA attacks using 100 000 traces. The power models have been chosen as i) an Sbox output bit and ii) a bit of the XOR result of an Sbox input and output. It is noteworthy that the attacks with common power models like Hamming weight (HW) models are also feasible, but not as efficient as those mentioned above.

We should stress here that although the AES accelerator is vulnerable to these state-of-the-art attacks, the effort an attacker needs to put in to recover the key is higher compared to e.g., the cases of Atmel's XMEGA [8] and KeeLoq [4]. This hardness results mainly from the low-power feature of the underlying technology. Further, as stated, we kept the MCU in an idle state to be able to observe the leakage of the AES module. The power consumption peaks related to the normal operation of the MCU are actually much higher than that of the AES module (see Fig. 4(a)). Such high power peaks are due to the FRAM reads (as stated followed automatically by a write) as the program code (instructions) has

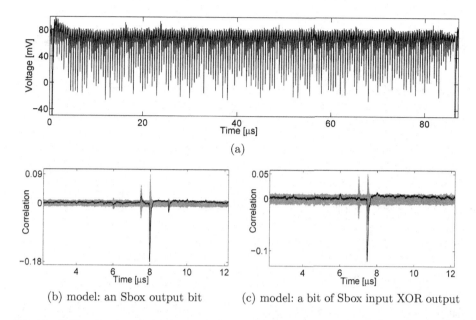

(a)

(b) model: an Sbox output bit   (c) model: a bit of Sbox input XOR output

Fig. 3. AES in hardware: (a) a sample trace (full AES), (b) and (c) CPA attack results using 100 000 traces

been stored in FRAM. In short, when the MCU operates simultaneously with the AES accelerator, the above-presented attacks become harder.

3.3 AES in Software

As a case study to examine the leakage of the normal operation of the MCU, we took the AES-128 implementation recommended by TI and publicly available at http://www.ti.com/tool/AES-128. Both encryption and decryption functions are written in C, and we used IAR Embedded Workbench IDE to compile the *encryption* code and ran it on our evaluation kit at a frequency of 8 MHz. We first realized that the length of the traces is not constant, and the implementation needs different number of clock cycles depending on the plaintext. The source of this issue was found in the way that *multiply by 2* (used for MixColumns) has been implemented:

```
unsigned char galois_mul2(unsigned char value)
{
    if (value>>7)
    {
        return ((value << 1)^0x1b);
    } else
        return (value << 1);}
```

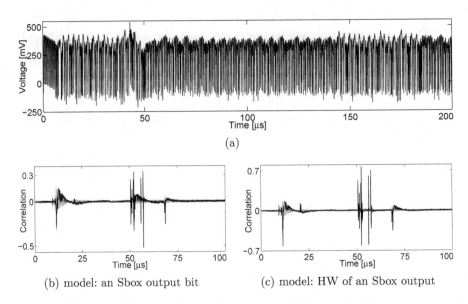

(a)

(b) model: an Sbox output bit   (c) model: HW of an Sbox output

Fig. 4. AES in software @ 8 MHz: (a) a sample trace (first round), (b) and (c) CPA attack results using 100 000 traces

As a result the implementation is vulnerable to a classical timing attack by predicting whether the extra reduction (XOR by the polynomial) is required or not. We have implemented such an attack and could easily recover the key using less than 5000 timing measurements (of the full encryption). Further, such an implementation is trivially vulnerable to a simple power analysis attack, where by observing each power trace the adversary can conclude whether the extra reduction was performed or not directly leading to a shrink in the key space.

Regardless of this issue we examined the efficiency of the start-of-the-art power analysis attacks. A sample trace covering the first round of the encryption is shown in Fig. 4(a). In short, several attacks by different hypothetical power models are feasible, as expected. The results of two CPA attacks predicting a bit of an Sbox output as well as the HW of the Sbox output are shown in Fig. 4. It is noteworthy that compared to the AES accelerator, the leakages associated with the execution of the MCU instructions are an order of magnitude easier to exploit. In fact, the ultra-low-power feature of the underlying MCU does not play an important role to harden the attacks.

3.4   Cache

As explained in Sect. 2, the FRAM is equipped with a cache to accelerate the access to consecutive memory locations when the MCU runs at a higher speed than the FRAM. In order to examine the effect of cache miss/hits we considered the AES encryption function in software (the case study of Sect. 3.3). To enable the cache we adjusted the clock source settings to operate the MCU at

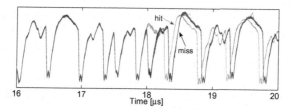

Fig. 5. AES in software @ 16 MHz: detection of cache miss/hit through power traces

a frequency of 16 MHz. Figure 5 shows a couple of traces during the SubBytes operation (one Sbox call). It can be seen that the traces belonging to cache hit and cache miss are clearly distinguishable. Therefore, a trace-driven cache attack is possible. In other words, by comparing a couple of traces the attacker would be able to detect whether each Sbox call caused a cache miss or not.

For an attack scenario let us consider two consecutive Sbox calls $S(p_1 \oplus k_1)$ and $S(p_2 \oplus k_2)$. If by observing the power traces the attacker detects a cache hit during the second Sbox call, it can be directly concluded that the two Sbox calls accessed nearby memory locations. Therefore, the attacker can gain certain information about $\Delta k = k_1 \oplus k_2 = p_1 \oplus p_2$. With respect to the underlying cache architecture, i.e., 64-bit lines (8 bytes), the five most significant bits of Δk can be recovered by the attacker. This is true, if the Sbox table starts at a location in FRAM corresponding to the first byte of a cache line. In other words, the first entry of the Sbox table needs to be stored in a location with address xx...xx000. Otherwise, the recovered bits of Δk is reduced to the four most significant bits.

As a proof of concept we developed a scenario to perform such an attack. In such a scenario we collected 256 power traces $T_{i \in \{0,...,255\}}$ where the first plaintext byte p_1 is constantly set to an arbitrary value, e.g., 0, and the second one $p_2 = i$. The rest of the plaintext bytes can be arbitrarily selected. By observing the power traces and detecting a cache hit with $p_2 = p_2'$, a part of $k_1 \oplus k_2 = p_2'$ is recovered. Indeed, it is not required to collect all 256 traces; once a cache hit is detected the process can be terminated. For the second phase of the attack, again at most 256 traces are collected with plaintext bytes $p_1 = 0$, $p_2 = p_2'$, and $p_3 = i$. The same process is repeated to find the colliding case for $p_3 = p_3'$ and recovering a part of $k_1 \oplus k_3 = p_3'$. The selection of $p_2 = p_2'$ is necessary to avoid replacing the part of the cache filled during the first Sbox call. This process is repeated for the other plaintext bytes. At the end of the attack, the key space is limited to $2^5 \cdot 2^{3 \times 16} = 2^{53}$ or to $2^4 \cdot 2^{4 \times 16} = 2^{68}$ depending on the location the Sbox table is stored in. However, the attack can be extended to the second round and recover more relations to again shrink the key space.

An important issue, which should be mentioned, is that since the program code is also stored in FRAM, the execution of the instructions (fetching them from FRAM) by the MCU also affects the cache misses/hits. In the above-given example all 16 Sbox calls are trivially performed in a loop. If this is not the case, the instructions performed between two consecutive Sbox calls can already replace the interesting line of the cache and avoid any cache hit by the second call. The presented attack scenario is only an example of a common scheme in the presence of a cache. It should also be emphasized that since the cache is shared between the program and data memory, exploiting the leakage by timing attacks (as a time-driven cache attack) is not trivial. In general, we show the leakage sources which should be taken into account when using such an MCU in a security-related application.

3.5   Internal Architecture

As the last case study we considered a masking scheme, which has been developed to provide security against side-channel attacks at any order. The scheme, which is based on the work presented in [14] has been implemented on an FRAM-based MCU as a proof of concept [7]. Therefore, we could easily integrate the same program code on our platform and perform the evaluations.

With respect to Fig. 6, we restate the underlying scheme. In a classical first-order Boolean masking (Fig. 6(a)), x and m (resp. input and random mask) are given to the device, which generates two outputs as $S(x \oplus m \oplus k)$ and q : $S(x \oplus m \oplus k) \oplus S(x \oplus k)$ as a shared representation of $S(x \oplus k)$. Such a scheme is certainly vulnerable to a second-order attack combining the leakages associated with the output shares. The concept followed in [7,14] is to involve more random data in the computations in such a way that the look-up tables g_1, R, and RC are precomputed based on the predefined key k and random data a_1, a_2, and a_3 in a secure environment (see Fig. 6(b)). During the operation (similar to the classical Boolean masking) x and m are given to the device, and all the operations are performed by the aforementioned look-up tables. As each of the look-up tables involves a random a_i which is independent of the others, the adversary should not be able to recover any information by combining the leakage of the look-up tables and/or the output shares.

After a random selection of a_1, a_2, and a_3, the precomputations (Algorithm 1) are supposed to be performed in a secure environment, i.e., no side-channel measurement is permitted. After finishing all operations in the operational phase, e.g., for an Sbox as shown in Algorithm 2, $g_2(\cdot, \cdot)$ can be applied on (s_1, s_2) to obtain the unmasked result (in this case $S(x \oplus k)$).

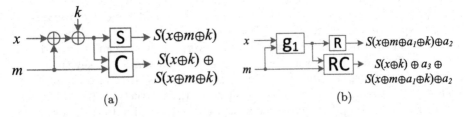

Fig. 6. (a) Classical Boolean masking, (b) The scheme of [7]

Algorithm 1. Look-up Table Precomputation

---

input : k, a_1, a_2, and a_3

output: $g_1(\cdot,\cdot)$, $R(\cdot)$, $RC(\cdot,\cdot)$, and $g_2(\cdot,\cdot)$

$\forall i,j; \; g_1(i,j) = \; i \oplus j \oplus a_1$

$\forall i; \quad R(i) = \quad S(i \oplus k) \oplus a_2$

$\forall i,j; \; RC(i,j) = S(i \oplus k) \oplus a_2 \oplus S(i \oplus j \oplus k \oplus a_1) \oplus a_3$

$\forall i,j; \; g_2(i,j) = \; i \oplus j \oplus a_3$

Algorithm 2. Operation

---

input : x, m, $g_1(\cdot,\cdot)$, $R(\cdot)$, and $RC(\cdot,\cdot)$

output: (s_1, s_2)

$g = \; g_1(x,m) \; ; \; /* \quad :x \oplus m \oplus a_1 \hspace{4cm} */$

$s_1 = R(g) \quad \; ; \; /* \quad :S(x \oplus m \oplus a_1 \oplus k) \oplus a_2 \hspace{2.3cm} */$

$s_2 = RC(g,m) \; ; \; /* \quad :S(x \oplus m \oplus a_1 \oplus k) \oplus a_2 \oplus S(x \oplus k) \oplus a_3 \; */$

A simplified and reduced version of the LED cipher [6] has been considered in [7] as an example to be implemented by the above-restated scheme. This reduced LED consists of only four rounds (cf. Fig. 7) and a 16-bit (4×4) data width (i.e., it works only on the first column of the full LED state). We integrated the corresponding code and ran the MCU at a frequency of 8 MHz. It should be noted that before each encryption the look-up tables g_1, g_2, R, and RC are recomputed, i.e., there is no mask reuse in the whole scheme. By means of appropriate trigger signals as well as inserting a large enough gap between the precomputation phase and the operational phase we made sure to measure the power consumption of the MCU only during the operational phase. As we planned to perform a *non-specific* t-test, we collected 150 000 traces while a fixed value (as 0) or a random one – in a randomly interleaved fashion – was given to the MCU as plaintext. The result of both first- and second-order univariate t-tests are shown in Fig. 8. Unexpectedly, the tests report both first- and second-order leakages.

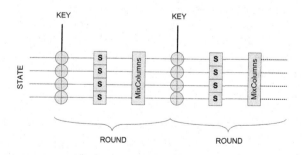

Fig. 7. Reduced LED (taken from [7])

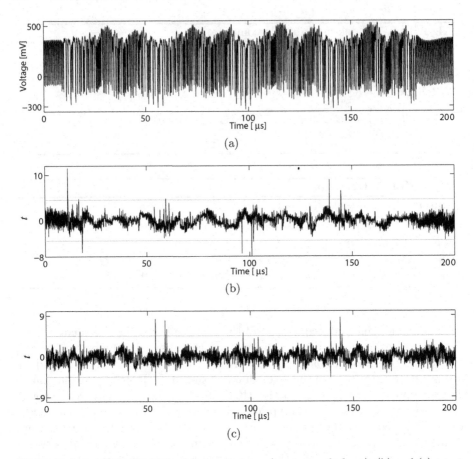

Fig. 8. Reduced LED @ 8 MHz: (a) sample trace (operational phase), (b) and (c) t-test results first- and second-order respectively using 150 000 traces

As stated, the scheme is supposed to provide resistance against the attacks at any order. The exploitable leakage, that we presented, is not due to the underlying scheme. In other words, we do not report any flaws in the algorithm or in the implementation of [7]. Instead, we show that even theoretically-sound countermeasures can fall into failure because of the internal architecture of the underlying platform. By slightly changing the program code and performing many measurements and analyses, we found out that the instructions which perform the table look-ups are the source of the observed leakages. More precisely, the exemplary instruction

```
         mov.b     @pointer , m0
```

which implements the call to the look-up table $RC(g, m)$ (see Fig. 6(b) and Algorithm 2) causes such a leakage. Since the details of the MCU architecture are not publicly available, the reason of the observed leakage cannot be easily pinpointed. Further, although we showed an exploitable leakage, the evaluation we performed (non-specific t-test) cannot give a clear assessment on the hardness of an attack exploiting such leakage.

We give an example to show the strong effect of the MCU's internal architecture on the side-channel vulnerability. We observed that if a random value is written to a location in FRAM the leakage associated with this write operation depends on the value which has been previously stored in that location. In other words, suppose that x has been stored at location **address**. A write operation, which stores a random value m at location **address**, leads to a leakage associated with the value of x as well. We observed such leakage during the evaluation of the above-expressed reduced LED implementation. At one point in the code (during the operational phase) a masked intermediate value is stored at a location where the unmasked plaintext had been stored before (at lines 4 and 7 of the below code):

```
1     mov        #STATE , pointer
      rlam       #4 , st0
3     add        st1 , st0
      mov.b      st0 , 0(pointer)
5     rlam       #4 , st2
      add        st3 , st2
7     mov.b      st2 , 1(pointer)
```

In order to avoid such a strong leakage (shown in Fig. 9) we cleared the contents of this location during the precomputation phase by:

```
1     mov        #STATE , pointer
      mov.b      #0x00 ,0(pointer)
3     mov.b      #0x00 ,1(pointer)
```

This indeed is an evidence to the statement given above. We believe that such leakage is due to the FRAM architecture as well as the way a write operation is performed.

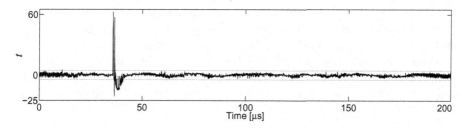

Fig. 9. Reduced LED (uncleared #STATE) @ 8 MHz: first-order t-test result using 35 000 traces

4   Conclusions

In this work we have extensively examined the side-channel vulnerability of FRAM-based MCUs of Texas Instruments as a platform for cryptographic applications. The motivation of this work is related to the relevant announcements dealing with the ultra-low-power feature of such MCUs and the claims on the hardness of power analysis attacks. Hence, we focused only on the power consumption of the underlying device and presented the corresponding evaluation results. The covered targets include the embedded AES accelerator (hardware), ordinary instructions of the MCU (software), FRAM cache, and the MCU internal architecture. In short, by means of practical investigations we confirm the hardness (but still feasibility) of the attacks on the embedded AES accelerator compared to similar targets of such attacks, e.g., the embedded AES core of Atmel's XMEGA MCUs. Such a hardness is mainly due to its low-power technology which leads to a high noise level in the measurements. However, when a cryptographic algorithm is implemented by the general-purpose MCU instructions, our practical results showed feasibility of straightforward and common DPA attacks without any serious difficulties. The cache, which has been integrated to accelerate the FRAM accesses, also comes with known security issues. Since an FRAM read must be followed by an FRAM write with the same value due to its destructive nature, an FRAM access consumes much more energy compared to a cache access. Hence, cache hit/miss can be clearly distinguished by observing the power traces. Although the cache is shared between the program and data memory (in MSP430FR5xxx family), we have shown that the trace-driven cache attacks (which exploit the sequence of cache misses/hits) are expectedly feasible. We also took a masking scheme into account, that has been developed in particular for platforms with a large non-volatile memory, e.g., FRAM-based MCUs. The scheme is based on precomputed randomized look-up tables and is expected to provide security against side-channel attacks of any order. Although there are no theoretical flaws in its developments, we have demonstrated that its implementation cannot pass a general leakage assessment test. The reason for such a failure lies in the details of the implementation platform (the MCU) regardless of the soundness of the underlying masking scheme.

On the one hand, the results we presented here are more or less expected as we targeted an unprotected platform where side-channel analysis should be feasible. On the other hand, this work gives an overview about the feasibility of exploiting various leakage sources of the underlying platform. Such information spreads awareness of the available leakage sources, and is certainly useful for cryptographic engineers, who deal with such a platform for security-related applications.

Acknowledgment. The authors would like to thank Stéphanie Kerckhof and François-Xavier Standaert from Université catholique de Louvain for their kindness in providing the source code of the masked implementation of the reduced LED of [7].

References

1. Balasch, J., Gierlichs, B., Verdult, R., Batina, L., Verbauwhede, I.: Power analysis of Atmel CryptoMemory – recovering keys from secure EEPROMs. In: Dunkelman, O. (ed.) CT-RSA 2012. LNCS, vol. 7178, pp. 19–34. Springer, Heidelberg (2012)
2. Bilgin, B., Gierlichs, B., Nikova, S., Nikov, V., Rijmen, V.: Higher-order threshold implementations. In: Sarkar, P., Iwata, T. (eds.) ASIACRYPT 2014, Part II. LNCS, vol. 8874, pp. 326–343. Springer, Heidelberg (2014)
3. Brier, E., Clavier, C., Olivier, F.: Correlation power analysis with a leakage model. In: Joye, M., Quisquater, J.-J. (eds.) CHES 2004. LNCS, vol. 3156, pp. 16–29. Springer, Heidelberg (2004)
4. Eisenbarth, T., Kasper, T., Moradi, A., Paar, C., Salmasizadeh, M., Shalmani, M.T.M.: On the power of power analysis in the real world: a complete break of the KeeLoq code hopping scheme. In: Wagner, D. (ed.) CRYPTO 2008. LNCS, vol. 5157, pp. 203–220. Springer, Heidelberg (2008)
5. Goodwill, G., Jun, B., Jaffe, J., Rohatgi, P.: A testing methodology for side-channel resistance validation. In: NIST Non-Invasive Attack Testing Workshop, Nara (2011)
6. Guo, J., Peyrin, T., Poschmann, A., Robshaw, M.: The LED Block Cipher. In: Preneel, B., Takagi, T. (eds.) CHES 2011. LNCS, vol. 6917, pp. 326–341. Springer, Heidelberg (2011)
7. Kerckhof, S., Standaert, F.-X., Peeters, E.: From new technologies to new solutions exploiting FRAM memories to enhance physical security. In: Francillon, A., Rohatgi, P. (eds.) CARDIS 2013. LNCS, vol. 8419, pp. 16–30. Springer, Heidelberg (2014)
8. Kizhvatov, I.: Side channel analysis of AVR XMEGA crypto engine. In: WESS 2009. ACM (2009)
9. Kocher, P.C., Jaffe, J., Jun, B.: Differential power analysis. In: Wiener, M. (ed.) CRYPTO 1999. LNCS, vol. 1666, pp. 388–397. Springer, Heidelberg (1999)
10. Microsemi. Security: Protect Your Intellectual Property. http://www.microsemi.com/products/fpga-soc/security
11. Moradi, A., Barenghi, A., Kasper, T., Paar, C.: On the vulnerability of FPGA bitstream encryption against power analysis attacks: extracting keys from xilinx Virtex-II FPGAs. In: CCS 2011, pp. 111–124. ACM (2011)
12. Moradi, A., Oswald, D., Paar, C., Swierczynski, P.: Side-channel attacks on the bitstream encryption mechanism of Altera Stratix II: facilitating black-box analysis using software reverse-engineering. In: FPGA 2013, pp. 91–100. ACM (2013)
13. Oswald, D., Paar, C.: Breaking mifare DESFire MF3ICD40: power analysis and templates in the real world. In: Preneel, B., Takagi, T. (eds.) CHES 2011. LNCS, vol. 6917, pp. 207–222. Springer, Heidelberg (2011)
14. Standaert, F.-X., Petit, C., Veyrat-Charvillon, N.: Masking with randomized look up tables. In: Naccache, D. (ed.) Cryphtography and Security: From Theory to Applications. LNCS, vol. 6805, pp. 283–299. Springer, Heidelberg (2012)
15. Texas Instruments. Introducing advanced security to low-power applications with FRAM-based MCUs (2012). www.ebv.com/fileadmin/design_solutions/php/download.php?path=uploads%2Ftx_highlightcampaign%2FWolverine_Security_Whitepaper_01.pdf

16. Texas Instruments. Embedded Processing & DSP Resource Guide (2014). http://www.ti.com.cn/cn/lit/sg/sprt285f/sprt285f.pdf
17. Texas Instruments. User's Guide: MSP430FR58xx, MSP430FR59xx, MSP430FR68xx, and MSP430FR69xx Family (2014). http://www.ti.com.cn/cn/lit/ug/slau367f/slau367f.pdf
18. Xilinx. UltraScale Architecture. http://www.xilinx.com/products/technology/ultrascale.html

Side Channel Attacks on Smartphones and Embedded Devices Using Standard Radio Equipment

Gabriel Goller[1](✉) and Georg Sigl[2]

[1] Giesecke and Devrient, Munich, Germany
gabriel.goller@gi-de.com
[2] Institute for Security in Information Technology,
Technische Universität München, Munich, Germany
sigl@tum.de

Abstract. Side Channel Attacks are a powerful instrument to break cryptographic algorithms by measuring physical quantities during the execution of these algorithms on electronic devices. In this paper, the electromagnetic emanations of smartphones and embedded devices will be used to extract secret keys of public key cryptosystems. This will be done using standard radio equipment in combination with far-field antennas. While such attacks have been shown previously, the details of how to find relevant emanations and the limits of the attack remain largely unknown. Therefore, this paper will present all the required steps to find emanations of devices, implement a side channel attack exploiting ultra high frequency emanations and discuss different test setups. The result is a test setup which enables an attacker to mount a side channel attack for less than 30 Euros.

1 Introduction

Side Channel Attacks (SCA) on processors of cryptographic algorithms, which are known for more than a decade now, are a very strong measure to break cryptographic algorithms. The basic idea of all side channel attacks is to measure a physical quantity of a processor during the processing of cryptographic algorithms and then extract information about the secrets of the algorithms out of these measurements. Such quantities can be the timing [1], the power consumption [2], electromagnetic emanations or even sound [3]. While most of these attacks require the attacker to have physical access to the device for these measurements (e.g. power consumption), recently also side channels were found where the attacker can make the measurements from a certain distance (e.g. sound, electromagnetic emanations, time). These attacks can be mounted

Gabriel Goller—This work has been partly supported by the German Bundesministerium für Bildung und Forschung as part of the project SIBASE with Förderkennzeichen 01IS13020E. Responsibility for the content of this publication lies with the authors.

© Springer International Publishing Switzerland 2015
S. Mangard and A.Y. Poschmann (Eds.): COSADE 2015, LNCS 9064, pp. 255–270, 2015.
DOI: 10.1007/978-3-319-21476-4_17

remotely, so that the attacked device need not be in the possession of the attacker.

A side channel which offers a lot of possibilities is the electromagnetic emanation of a device. This is especially true for modern devices, with processors that run at frequencies in the high MHz or even GHz range. At these frequencies, each signal line which carries such high frequency components can act as an antenna, and possibly emanate secrets which can then be measured using antennas or near-field probes. This can be wires connected to the processor, or even signal lines inside the processor [5].

In this paper, it shall be researched if such emanations can be measured using standard radio equipment. The sensors connected to the radio receivers are primarily far-field antennas, but also near-field probes are considered. Although primarily a smartphone will be used during the experiments, other smartphones as well as single-board computers will be examined, too.

The possibility of side channel attacks using electromagnetic radiation, or more generally the possibility that circuits emanate high-frequency signals that possibly leak secret information is already known since 1982, when the NSA TEMPEST program internally published the "NACSIM 5000" handbook [4]. This classified handbook gives advice on the design of devices which are shielded against such attacks and describes the attacks themselves. The documents were released in December 2000, which led to numerous publications in the field of side channel attacks.

In 2003, Agrawal et al. published a comprehensive paper about the possibility to use electromagnetic leakage for side channel attacks (see [5]). They evaluated the emanations of several devices, including smartcards and a PCI bus based SSL accelerator. They found emanations using near-field probes as well as a far-field log-periodic antenna. Most of these signals were on frequencies which were harmonics of the clock frequency of the evaluated devices.

Aboulkassimi et al. showed in 2011 [6] and 2013 [7] that it is possible to extract the key of an AES encryption executed on a Java-based mobile phone using an electromagnetic near-field probe. They used a differential side channel analysis approach and their attack succeeded with only 250 traces.

In 2013, Montminy et al. [8] succeeded in extracting the key of an AES encryption running on a 32-bit processor with a clock frequency of 50 MHz using a setup consisting of a near-field probe and a software defined radio. Using a differential side channel attack, they were able to extract all keybits using 100000 traces.

In 2012, Jun et al. and Kenworthy et al. introduced the possibility to attack smartphones and tablets using near-field and far-field probes in combination with radio receivers, software defined radios and oscilloscopes. They succeeded in extracting keys of RSA and ECC and showed the possibility of a differential side channel attack against AES (see [9–11]).

This paper will introduce several approaches for mounting a side channel attack using standard radio equipment with different types of test setups. The different components of the test setups are introduced in Sect. 2, which is followed by Sect. 3 where we explain how to find the right frequencies where signals are emanated. In Sect. 4, a practical attack on a smartphone using a far-field antenna is implemented, evaluated and also tested with different devices. This is followed by Sect. 5, where a very low-cost setup for mounting the presented attack is evaluated.

2   Experimental Setup

In general, the different hardware-setups used for finding emanations and conducting the attack are based on the same layout (see Fig. 1). This layout consists of an antenna/near-field probe to receive the signal, a radio device which translates the analog signals into digital samples and a laptop which is responsible for signal and data processing. This makes it possible to scan a wide range of frequencies, use different antennas or sensors for different experiments, quickly change parts of the signal processing structure and save data for later analysis.

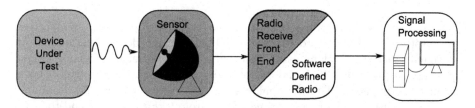

Fig. 1. Layout of the test setup (gray elements signify analog parts of the setup)

2.1   Sensors

Depending on the experiment, different antennas with different characteristics have to be used. To evaluate the emanations of the device under test (DUT), the antenna has to offer a high antenna gain over a large bandwidth, which can be achieved with a log-periodic antenna such as the HyperLOG 4025 by Aaronia AG. For the evaluation of the found emanations, the antenna should have a high antenna gain, but the bandwidth only needs to cover the found emanations. A high-gain-narrow-bandwidth antenna like a bi-quad type offer a good performance for these tasks, and can be built with very basic components (Fig. 2).

Alternatively, near-field probes can be used to capture emanations. While the far-field antennas in general have to be specifically designed for a certain bandwidth, these probes work over a very large range of frequencies. However, they have to be placed very near to the device, and the emanations can only be measured if the probe is directly above the source. Since this paper is focused on attacks using far-field antennas and at the frequencies at which the device under test is running, a near-field probe was only used to find emanations. The probe used was a Langer ICR HV 150-27 with a frequency range from 1.5 MHz to 6 GHz.

Fig. 2. Different sensors: Bi-Quad antenna and Langer ICR HV 150-27

2.2 Radio Device

There are two devices responsible for the radio reception, an analog radio receiver and a software defined radio (SDR). A SDR is a radio device where most of the signal processing is done using digital algorithms rather than analog filters. By combining it with a high-end analog radio receiver, the advantages of both systems can be taken. For the analog part, a ESN test receiver by Rohde & Schwarz was used. It offers a frequency range from 9 kHz to 1 GHz and tools to analyze the signals within this spectrum. To combine this receiver with a digital signal processing system, a USRP N210 software defined radio was connected to the IF output of this receiver. The N210 is able to sample an analog signal with 100 MSps with a resolution of 14 Bit. With this measuring system, the high bandwidth of the test receiver can be combined with the advantages of a digital signal processing system. To further increase the signal strength, a PA 303 30 dB preamplifier from Langer was inserted between the antenna and the test receiver.

An alternative reception system (presented and only used in Sect. 5) uses a standard DVB-T stick from Gixa Technology as an alternative to the ESN, the USRP and the preamplifier from above, which drastically reduces the price as well as the size of the measuring equipment, even compared to a setup for measuring power consumption. Furthermore, since no alteration of the hardware of the DUT is required and most of the signal processing is happening in software, it is possible to mount an attack even with little knowledge about hardware designs or measurement engineering. Internally, the DVB-T stick consists of two chips which roughly do the same job as the ESN radio receiver and the USRP software defined radio. A R820T chip by Rafael Microelectronic is used to tune in and downconvert a radio signal, which is then converted to a digital I/Q signal by a RTL2832U chip by Realtek. There exist numerous DVB-T sticks with this hardware combination, and a list of compatible sticks can be found at the project website [15].

Both systems can be seen in Fig. 3, which also shows the far smaller size of the DVB-T stick compared to the ESN-USRP combination.

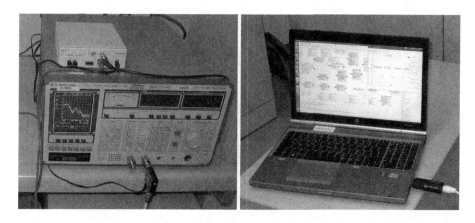

Fig. 3. ESN test receiver with USRP N210 and preamplifier (left), compared to the DVB-T stick (plugged into Laptop, right)

2.3   Software Components

To process the digital signals, the open source software GNU Radio was used. This makes it possible to test whether relevant signals are emanated, be it directly or in a modulated way. Further processing was done using GNU Octave, an open source tool for numerical computations [12].

2.4   Device Under Test (DUT)

Though in Sect. 4.4 several other devices will be examined, the primary research was conducted on an Android-based smartphone. The only alteration of the hardware was to remove a shielding plate above the main circuit to make it easier to find and measure emanations (see Fig. 4). However, in Sect. 4.3 the attack will also be tested with the shielding plate. Software-wise, the system was rooted to be able to influence the CPU clock frequency. The app which computes the cryptographic algorithms was written in Java, with the cryptographic parts written in C using the Android Native Development Kit.

Positioning of Near-Field Probes. To get good results with the near-field probes, the probe has to be near the source of the emanations. Therefore the first step is to find a position where emanations take place. Such a position can be found by connecting the probe to a oscilloscope, then produce a processor load and change the position of the probe until a signal is received. By doing this, a position was found directly above a capacitor next to the main CPU where a signal is emitted by the smartphone (see Fig. 4).

2.5   Software on DUT

To evaluate the possibility of a side channel attack, a square and multiply algorithm was implemented on the device. This algorithm, which is one of the stan-

Fig. 4. DUT, with shielding plate (red frame) and position where the near-field probe captured signals (red cross) (color figure online).

dard algorithms for implementations of RSA, can be used to calculate the result of the equation

$$m = c^d \bmod N, \tag{1}$$

with m being the decrypted message, c the encrypted message, d the secret key and N a publicly known integer. The algorithm can be described by the following pseudo-code:

```
function square-and-multiply(Number c, Integer d, Modulus N)
    result = 1
    for each bit(d) from (number_of_bits(d) - 1) downto 0
        result = square(result) mod N //square operation
        if bit(d) == 1
            result = (c * result) mod N //multiply operation
        end if
    end for
    return result
end function
```

For the algorithm, the OpenSSL library was used [14]. This was done in such a way that the algorithm itself was a custom C implementation, but the square and the multiply operations were taken from the OpenSSL library.

3   Emanations of Smartphones

The first challenge when trying to implement a side channel attack is to find appropriate signals which contain information correlated to the activity of the processor of the device. This can be very difficult when the sensor is a far-field antenna, because there are many other signals caused by terrestrial radio stations, such as mobile phone networks, DVB-T and others. Therefore, it is easier to use a near-field probe as a sensor to find the emanations of a device. During research, 3 approaches to find the relevant frequencies where the smartphone emanates signals were developed, one using an antenna, one using a near-field-probe and one by making an educated guess.

The first approach was done with a wide-band antenna and consists of filtering the external disturbances from the signal, so that only the signals emanated by the DUT remain. The idea is that the signal s measured by the antenna at a certain frequency consists of two components: The signals emanated by the DUT s_{DUT} and the signal emanated by other sources s_{others}, which can be written as

$$s = s_{DUT} + s_{others}. \tag{2}$$

In this equation, s and s_{others} can be measured by measuring two times in a row, one time with and one time without the DUT enabled. That way, s_{DUT} can be calculated by computing

$$s_{DUT} = s - s_{others}. \tag{3}$$

While it is not possible to filter all the external signals using this system, because s_{others} is not constant, the search space is reduced to a few frequencies at which the DUT possibly emanates signals, which can then be checked manually. The results of this approach for a frequency range from 400 MHz to 1 GHz can be seen in Fig. 5.

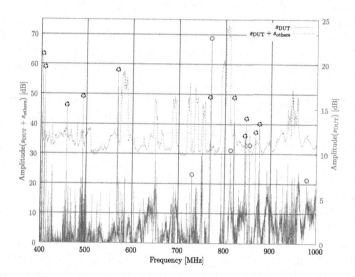

Fig. 5. Results of the antenna measurements: received signal strength of the DUT and disturbances (blue, dashed line), calculated emanations of only the DUT (red, solid line). Signals really emanated by the DUT are marked with a green circle, false positives are marked with a dashed black circle (Color figure online)

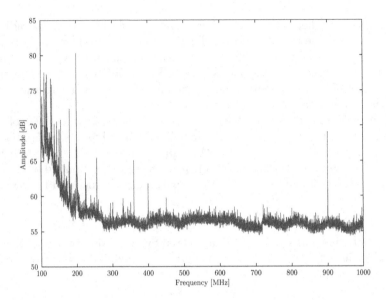

Fig. 6. Results of the near-field measurements: Emanations on several frequencies, especially 900 MHz, the CPU clock frequency.

The second approach is done by measuring with a small near-field probe directly above the main processor. The near-field probe is not affected by the disturbing radio signals, and thus the measured signals are directly emanated by the smartphone. Since the emanations found by the near-field probe are only magnetic fields, it is necessary to check in a second step whether the emanations can also be measured with the far-field antenna. During this experiment, the clock frequency of the smartphone was set to a fixed frequency of 900 MHz. As it can be seen in Fig. 6, at this frequency signals were emanated by the device.

The third approach is done by making an educated guess to find the relevant frequencies. This can be done without any sensors by studying the manual of the device to find out which clock frequencies exist. Since the emanations are caused by coupling of high frequency signal generators with other parts of the circuit [4], there is a high probability that there are signals emanated on the frequencies used by the active high-frequency elements of a smartphone, e.g. clock generators.

Using these 3 approaches, several signals emanated by the smartphone were found. These signals could be categorized into signals emanated by the main processor and, far stronger, signals emanated by the display. The signals emanated by the display are only measurable when the touchscreen is turned on, while the signals emanated by the main processor can be always measured when the processor is doing work. However, to be able to capture the processor signals, the smartphone has to be configured so that the CPU clock is kept at a distinct rate, because otherwise the frequency is changed depending on the workload. This is also the reason why the far-field antenna did not receive a signal at 900 MHz, because during this experiment the device was not running at a fixed clock frequency.

4   Side Channel Attack Using Far-Field-Antennas

4.1   Correlation to Computations

To successfully mount a side channel attack, it is crucial to find out which of the signals contain information about the main processor. To do this, during the measurements, a periodical workload is executed. This way, if there is a correlation of the processor load with the signals, the emanations should also contain a periodical component. Doing this, the two categories of signals were analyzed, which led to two different results.

It is not possible to extract information of the main processor from the signals emanated by the display. However, these signals contain information about changes in the display content and the state of touchscreen. In a first experiment, it was possible to measure the blinking of a cursor and the signal when a finger touches the display from a distance as far as 3 m. However, since the focus of this paper are side channel attacks on cryptographic algorithms, these signals were not further considered.

The emanations of the main processor however are correlated with the data and instructions processed there. Every time the processor is working, the amplitude of the signals is higher than in an idle mode, where the amplitude goes down to nearly zero (e.g. during the call of the standard C-function "sleep()"). To check

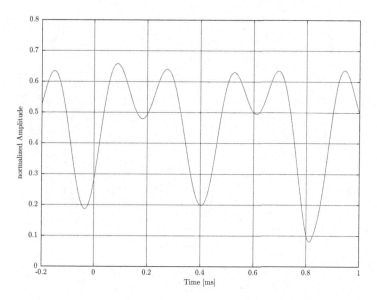

Fig. 7. A square (0–0.4 ms) and a multiply (0.4–0.8 ms) operation, not distinguishable.

if this behavior can be used for a side channel attack, the signal was evaluated during the execution of the Square-and-Multiply algorithm from Sect. 2.

4.2 Square-and-Multiply Algorithm

Using OpenSSL, a Square-and-Multiply algorithm was implemented. If it is possible to distinguish square and multiply operations, the secret exponent d can be extracted during calculation of Eq. 1.

By isolating the single operations from each other with a *sleep()*-function, it can be checked if they cause different signals and can be distinguished that way. As it can be seen in Fig. 7, this is not the case. The operations can not be distinguished on first sight, and further investigations showed that this is not even possible by using statistical tools like the cross-correlation of the signal. This is probably due to the low sampling rate and the low signal-to-noise ratio.

When the *sleep()*-functions are removed, the single signals of the operations melt into a big block with a high amplitude, and thus the operations can neither be seen nor distinguished in the final signal. However, this can be improved by a common technique from the field of signal processing, which is to record the same signal several times in a row and then compute the average of these recorded signals [13, page 367]. This way, it is possible to extract the key, because the signal of a 1-bit (square and multiply) can be distinguished from the calculation of a 0-bit (square only). This can also be seen in Fig. 8.

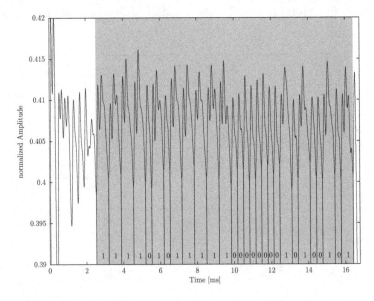

Fig. 8. Average of 1063 traces of a Square-and-Multiply algorithm (execution marked gray, key: 1010 0101 0000 0000 1111 1010 1111, computed backwards).

4.3 Evaluation of Attack

After proving that an attack using a far-field antenna is possible, a few parameters of the attack shall be evaluated here.

Number of Traces. Since for a successful attack it is necessary to average several traces (with a single trace being the recorded curve of a single execution of the algorithm), it is convenient to find the minimal number of traces for a successful attack. The challenge is however that the result of the attack is not computed by an algorithm, but by a human which interprets the results, which is highly subjective. To find an objective measure for the minimum number of traces, a property of the averaging was used. Since the average of many traces converges to a unique solution where the keybits can be extracted, an objective measure is to compare the average of i traces with the converged solution, and then see how much alike the curves are. A tool for such an analysis is the correlation coefficient, which can be used to compare two digital waveforms x and y with each other. It is defined as

$$\text{corr}(x, y) = \frac{\text{cov}(x, y)}{\text{std}(x)\text{std}(y)}, \tag{4}$$

where x and y are the waveforms to compare, $\text{std}(\cdot)$ is the standard deviation and $\text{cov}(x, y)$ is the sample covariance of the waveforms x and y defined as

$$\text{cov}(x, y) = \frac{1}{n - 1} \sum_{j=1}^{n} (x_j - \mu_x)(y_j - \mu_y), \tag{5}$$

with μ as the mean value of a waveform, x_j and y_j as the j–th element of the waveforms x and y and n as the total number of elements of the waveforms. Using these definitions, the following formula was used to calculate the correlation between the average of $i = 1 \ldots q$ traces t_i with the converged solution, which is the average of q traces:

$$y(i) = \text{corr}(\text{mean}(t_1, t_2, \ldots, t_i), \text{mean}(t_1, t_2, \ldots, t_q)). \tag{6}$$

$\text{mean}(\cdot)$ computes the average curve c of multiple traces t. Each point c_j of the curve c is defined by

$$c_j = \frac{1}{i} \sum_{p=1}^{i} t_{p,j} \tag{7}$$

with $t_{p,j}$ being the j-th element of the p-th trace t_p.

As it can be seen in Fig. 9, the correlation is already higher than 0.999 when more than 170 traces are averaged. This means that with 170 traces it should still be possible to extract the key, which was also confirmed by experimental results.

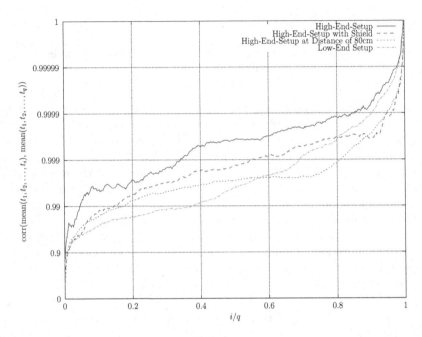

Fig. 9. Correlation of the average of i traces with the average of q traces ($q = 1894$ for high-end setup at distance of 80 cm and $q = 500$ for the other experiments.)

Maximum Distance. So far, the device was placed directly in front of the antenna. Increasing the distance decreases the quality of the signal in several ways. Obviously, the amplitude of the signal is reduced, which worsens the signal-to-noise ratio. Subsequently, the synchronization of the traces gets more difficult, which means that more traces are needed and even with more traces, it is harder to identify the different operations in the signal. However, it is still possible to extract the key from a distance of 80 cm using 1894 traces, but it is very hard to identify the different operations, even when comparing the curve with the result of the attack from a distance of 2 cm (see Fig. 10). As it can be seen in Fig. 9, when applying the correlation experiment from above to the data acquired at a distance of 80 cm, the average of 1530 traces is needed to get a correlation of 0.999 with the average curve of the 1894 traces shown in Fig. 10.

Shielding Plate. In the factory state the smartphone is equipped with a shielding plate which resides directly above the main circuits of the device. Since this does decrease the emanations of the device, it was removed so far. When putting it back into the device, the results are comparable to increasing the distance. The amplitude drops, resulting in a lower signal-to-noise ratio and thus requiring more traces than before. However, it was still possible to extract the secret of a Square-and-Multiply algorithm by averaging multiple curves. The maximum distance with the shielding plate is however drastically reduced, so that the

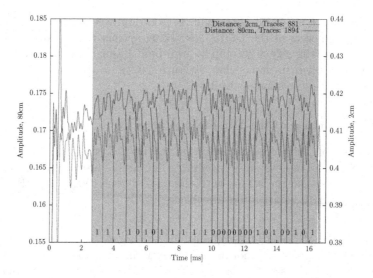

Fig. 10. Average of 1894 traces recorded at a distance of 80 cm.

antenna has to be directly next to the device during the attack. When applying the correlation experiment from above to 500 traces collected with the shielding plate installed and the antenna at a distance of 2 cm, it takes the average of 276 traces instead of 170 to reach a correlation of 0.999.

4.4   Other Devices

At first, the research was only conducted with a single smartphone. Since many more devices are based on the same processor architecture, it could well be that all these devices are vulnerable to the attack. Therefore, it was tested with different devices, not only in the area of smartphones, but also on single board computers. The results were that on all tested devices (3 smartphones and 2 single board computers) emanations were measurable at the clock frequency, and the attack could be performed successfully. While it was necessary to remove a shielding plate on two of the smartphones, this was not the case for one smartphone. Because the attack can also be mounted on single-board computers like the Raspberry Pi or the BeagleBone Black, it is unlikely that the phone-specific circuits (e.g. the antennas) are the reason for the emanations. Altogether, the results suggest that the emanations are not caused by an individual flaw in the circuit- or processor-design and could possibly affect many more devices.

5   A Low-Cost Setup for EM Analysis

Recently, a cheap alternative to radio receivers was introduced by making it possible to use DVB-T sticks as Software Defined Radios [15]. The sticks that can be used for this purpose offer a frequency range between 24 and 1850 MHz, a sampling rate of up to 2.5 MSps and a resolution of 8 Bit. While this is well below the parameters of the USRP, a working system could reduce the parts needed to mount the attack to only an own-built bi-quad antenna and a DVB-T stick, with total costs well below 30 Euros. Because of this, the attack was implemented with this setup as well.

5.1 Reproduction of the Far-Field Attack

The primary goal is to find out whether the DVB-T stick can compete with the performance of the system from above. The results are very promising: The attack works just like the attack with the system consisting of USRP and test receiver, however there are a few drawbacks.

Quality of the Signal. The quality of the signal is drastically reduced. Although both signals are normalized (which means that the amplitude is between 0 and 1), the signal is one magnitude smaller (0.4 vs. 0.016) than with the high-end setup. However, the amplitude of the noise is also much smaller when measuring with the DVB-T stick compared to the setup from above (see Fig. 11). To compare both measurements, the amplitude of a signal has to be compared with the amplitude of the noise, which can be done by computing the

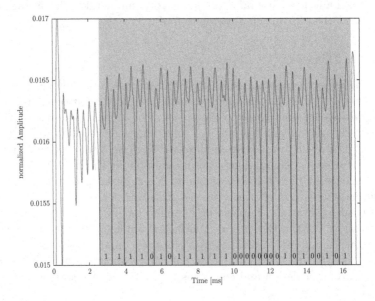

Fig. 11. Average of 1228 traces of a Square-and-Multiply algorithm recorded with a DVB-T stick (execution marked gray, key: 1010 0101 0000 0000 1111 1010 1111, computed backwards).

signal-to-noise ratio [16]. It can be estimated as

$$\text{SNR}_{\text{dB}} = 20\log_{10}\left(\frac{A_{\text{signal}}}{A_{\text{noise}}}\right), \tag{8}$$

where A_{noise} and A_{signal} are the root mean square amplitudes of the signal and the noise, respectively. This gives 11.82 dB for the DVB-T stick and 13.94 dB for the high-end system. This suggests that the low-end setup is not performing as bad as the small amplitude of the signal would suggest. The best comparison of the two systems however is the number of traces needed for a successful attack. Using the correlation technique from Sect. 4.3, the DVB-T stick can be compared to the system from above, which can be seen in Fig. 9. As it can be seen, to get the same correlation as above (0.999), twice as many curves are needed (346 vs. 170).

Maximum Distance. Another drawback is that the maximum distance is reduced due to the bad signal. Instead of 80 cm maximum distance with the other system, the low-cost system is not able to receive a signal when the distance is larger than ∼10 cm. Altogether, for the far-field attack the system does not perform as well as the original system, but the reduction of costs makes it a very good alternative, especially when the number of traces and the distance is not of vital importance.

6   Summary

In this paper, the already existing results on side-channel attacks on smartphones and embedded devices using electromagnetic emanations were further researched. This includes all parts of the experiment, starting with the search for emanated signals, continuing with the description and evaluation of the possible attacks and concluding with the development of a very low-cost test setup. It was shown that using a far-field antenna, it is possible to extract the secret key of a Square-and-Multiply algorithm by averaging several traces. While a lower distance is advantageous, the attack can also be conducted from a distance much larger than it is possible with the state of the art near-field probes. Finally, a very low-cost setup was implemented, which makes it possible to mount the attack with costs lower than 30 Euros, using a DVB-T stick and a self-built antenna. This enables even attackers with a very small budget to attack smartphones and embedded devices. The results of this paper show that hardware and software countermeasures have to be implemented in smartphones and embedded devices, or secure elements should be used for these cryptographic computations, especially since devices like smartphones and single-board computers tend to be used for more and more applications where security is vital, e.g. payment applications (smartphones) or industrial applications (embedded single-board computers).

References

1. Kocher, P.C.: Timing attacks on implementations of diffie-hellman, RSA, DSS, and other systems. In: Koblitz, N. (ed.) CRYPTO 1996. LNCS, vol. 1109, pp. 104–113. Springer, Heidelberg (1996)
2. Kocher, P.C., Jaffe, J., Jun, B.: Differential power analysis. In: Wiener, M. (ed.) CRYPTO 1999. LNCS, vol. 1666, pp. 388–397. Springer, Heidelberg (1999)
3. Genkin, D., Shamir, A., Tromer, E.: RSA key extraction via low-bandwidth acoustic cryptanalysis. IACR Cryptol. ePrint Archive **2013**, 857 (2013)
4. National Security Agency: NACSIM 5000 Tempest Fundamentals. Partially released in December 2000–February 1982
5. Agrawal, D., Archambeault, B., Rao, J., Rohatgi, P.: The EM side-channel(s). In: Kaliski, B.S., Koç, Ç.K., Paar, C. (eds.) CHES 2002, pp. 29–45. Springer, Heidelberg (2003)
6. Aboulkassimi, D., Agoyan, M., Freund, L., Fournier, J., Robisson, B., Tria, A.: Electromagnetic analysis (EMA) of software AES on java mobile phones. In: 2011 IEEE International Workshop on Information Forensics and Security (WIFS), pp. 1–6. IEEE (2011)
7. Aboulkassimi, D., Fournier, J., Freund, L., Robisson, B., Tria, A.: EMA as a physical method for extracting secret data from mobile phones. Int. J. Comput. Sci. Appl. (IJCSA) **2**(1), 16–25 (2013)
8. Montminy, D., Baldwin, R., Temple, M., Oxley, M.: Differential electromagnetic attacks on a 32-bit microprocessor using software defined radios. IEEE Trans. Inf. Forensics Secur. **8**(12), 2101–2114 (2013)
9. Kenworthy, G., Rohatgi, P.: Mobile device security: the case for side channel resistance. In: Proceedings of the 2012 Mobile Security Technologies Conference, California, USA (2012)
10. Jun, B., Kenworthy, G.: Is your mobile device radiating keys? Presentation, held at RSA Conference (2012)
11. Kenworthy, G., Rohatgi, P.: Mobile device security: the case for side channel resistance. Presentation, held at Mobile Security Technologies Workshop (2012)
12. Eaton, J., Bateman, D., Hauberg, S., Wehbring, R.: GNU Octave Free Your Numbers edition 3 for octave version 3.8.0 edition (2011)
13. Swanson, D.C.: Signal Processing for Intelligent Sensor Systems with MATLAB, 2nd edn. Taylor & Francis, Boca Raton (2012)
14. The OpenSSL Project: OpenSSL: The Open Source Toolkit for SSL/TLS. http://www.openssl.org. Accessed December 2014
15. Wiki, http://rtlsdr.org. Accessed December 2014
16. Johnson, D.H.: Signal-to-noise ratio. Scholarpedia **1**(12), 2088 (2006)

Author Index

Printed in the United States
By Bookmasters